FISHERY CO-MANAGEMENT
A Practical Handbook

Contents

Foreword

It is estimated that there are 51 million fishermen in the world of whom 50 million are small-scale, subsistence, or artisanal operators mainly living and working in the developing world. Unfortunately, the common-property fish resources that these fishermen depend upon for their livelihood are in a precarious state. Why do we find ourselves in this critical situation and what can we do to correct it? Researchers and stakeholders are actively searching for new forms of fishery management, and one of the most promising forms centres around approaches involving the users – the fishermen themselves – working in some form of shared or collaborative management with government authorities.

For many years, Canada's International Development Research Centre (IDRC) has maintained an active portfolio of projects examining co-management and community-based management in fisheries and other resource systems. Since the publication of *Managing Small-scale Fisheries* (Berkes *et al.*, 2001), there has been an increasing demand for guidance on what IDRC has learned about co-management, particularly across different geographical settings, socio-economic conditions, and histories of operation; and how it could apply to other types of fishing, link to other livelihoods, relate to other dynamic processes (such as the migration of fishermen), and respond to the seasonal nature of fish resources. This book attempts to respond to this demand by compiling recent experience from as wide a cross section of research as possible.

During the development of this book, both IDRC and the authors wrestled with the concept of co-management. Given the evolving nature of this science, for example, what does co-management cover and how widely is the concept accepted? Importantly, there has been increasing acceptance of the idea that co-management is not an end point but rather a process – a process of adaptive learning. Recognizing the diversity of both local contexts (ecological and social) and factors depleting the fishery (such and overfishing and habitat destruction), however, would it even be possible to put together a book of lessons learned? As you will soon discover, IDRC and the authors felt that it was neither possible nor desirable to produce a blueprint for fishery co-management. Rather, we agreed that it would be more useful to document the co-management process, as undertaken by both IDRC partners and others, and to put this experience into a form that could be shared with anyone interested in learning more about co-management and what others have learned. This shared and adaptive approach to learning is what this book is all about. In the pages that follow, you will find a complete picture of the co-management process: strengths, weaknesses, methods, activities, checklists and so on.

IDRC will continue to field-test this publication. Our evolving support for research on the co-management of natural resources is part of IDRC's ongoing work around governance, decentralization and adaptive learning, most of which are managed through the Centre's Rural Poverty and Environment programme. Details on this and other IDRC programmes can be found on our website: www.idrc.ca.

F. Brian Davy
Senior Program Specialist
Environment and Natural Resource Management
International Development Research Centre
Ottawa, Canada
email: bdavy@idrc.ca

Preface

In travelling around the world working on fisheries co-management, the question that people always seem to ask is: 'I understand the concept of co-management and it seems to be a good management strategy, but how do we make it work in practice?' This question was really the driving force behind the writing of this handbook. There seemed to be a need to have a practical reference on a process for community-based co-management. While there were a number of very good publications available which discuss individual activities and components of the process of community-based co-management, there was no one publication which could serve as a reference to the process from beginning or pre-implementation, through implementation, to post-implementation.

This handbook is meant to be a companion volume to the book, *Managing Small-scale Fisheries: Alternative Directions and Methods*, authored by Fikret Berkes, Robin Mahon, Patrick McConney, Richard Pollnac and Robert Pomeroy (2001). It is meant to provide more detail on the application of community-based co-management in small-scale fisheries. This handbook focuses on small-scale fisheries (freshwater, floodplain, estuarine or marine) in developing countries. The community-based co-management process described may also be relevant to small-scale fisheries in developed countries, as well as to other coastal resource (i.e. coral reefs, mangroves, seagrass, wetlands) management.

This handbook is based on years of experience conducting research on and practising community-based co-management in Asia, Africa, and the Caribbean and Latin America. There are far too many people who have shared their knowledge, experience and life with us to thank each one individually.

We hope this handbook will serve its purpose. We request feedback from you, the users, on how to improve the handbook. As co-management is a process, so is the development of a handbook. There is still a great deal to learn and share as we use community-based co-management to hopefully improve fisheries and marine resources and the lives of the people who depend upon these resources.

Robert S. Pomeroy and Rebecca Rivera-Guieb

Acknowledgements

The authors would like to thank the International Development Research Centre, and especially Brian Davy, for supporting this project. We would also like to thank the Oak Foundation, and especially Kristian Parker, for supporting in part, Robert Pomeroy's contribution to the project.

The authors would also like to thank the reviewers who provided useful comments for the writing and revisions of this handbook, including Alan White, Maria Hauck, Merle Sowman, Mafaniso Hara, Hugh Govan, John Kurien, Toby Carson, Rene Agbayani, Peter van der Heijden, Gary Newkirk, Patrick McKinney, Gaynor Tanyang, Elmer Ferrer, Mike Reynaldo, Dick Balderrama, Jean Harris, Yves Renard and John Kearney. Thank you also to Bill Carman of IDRC for shepherding this book through the publication process and to Jenny Thorp for her editing.

Acknowledgements are made to the University of Connecticut and Connecticut Sea Grant Program, especially Ed Monahan and Emilio Pagoulatos.

Robert S. Pomeroy and Rebecca Rivera-Guieb

List of Figures

List of Boxes

List of Tables

I Introduction

1   Why This Handbook?

IDRC, R. Ramlochand.

During the last decade, there has been a shift in the governance of fisheries to a broader approach that recognizes fishers' participation, local stewardship, and shared decision-making in the management of fisheries. Through this process, fishers are empowered to become active members of the fisheries management team, balancing rights and responsibilities, and working in partnership, rather than antagonistically, with government. This approach is called co-management.

Co-management is one of a number of promising new and alternative management approaches that have emerged in recent years for fisheries management. These new management approaches are part of a discussion of new directions for fisheries management presented in a recent book authored by Berkes *et al.* (2001), a companion volume to this publication. The authors present a vision for small-scale fisheries that sees the linkages between human and natural systems and recognizes the need for management approaches that address these linkages. It is a vision with a human face and a people focus – fishers and fishing communities. It recognizes that the underlying causes of fisheries resource overexploitation and environmental degradation are often of social, economic, institutional and/or political origins. It recognizes that fisheries management should focus on people, not fish, *per se.*

It is becoming increasingly clear that governments, with their finite resources, cannot solve all fishery problems. Local communities will need to take more responsibility for solving local problems. In order to do this, however, communities must be empowered and resources provided to make decisions locally and to take actions that meet local opportunities and problems. The assistance and support of government will still be needed to achieve these results, although the role and responsibilities of government will also need to change.

It is important for the fisheries manager to be creative and innovative. There is no blueprint formula for managing a fishery; each one is different. Different approaches will need to be tried and integrated. There will be success and there will be failure. There must be learning and adaptation. The community of fishers and the government, through a co-management arrangement, will need to work together to decide the best combination of approaches for their situation.

The concept of co-management has gained acceptance among governments, development agencies and development practitioners as an alternative fisheries management strategy to the top-down, centralized government management approach. However, the actual process of co-management has often been problematic as the definition of co-management is quite broad and means different things to different people. In addition, the complexity of issues and relationships and the time involved in planning and implementing co-management arrangements has also led to problems with implementation. This has often led to misunderstandings among the co-management partners and vague adoption of the concept as a fisheries management strategy.

1.1. What This Handbook Is

One of the difficulties in planning and implementing co-management is the lack of specific direction on 'how to do it'. There are a number of different activities and interventions in the process of community-based co-management. While there are a number of very good publications (many often difficult to obtain, especially in a developing country) which discuss individual activities and components of the process of community-based co-management, there is no single publication which provides a reference to the process from beginning or pre-implementation, through to implementation, and to turnover to the community or post-implementation. That is the purpose of this handbook; to provide a practical reference on a process for community-based co-management for use by the various co-management partners. This handbook is meant to be a working document that is revised and adapted to a specific situation.

This handbook focuses on small-scale fisheries (freshwater, floodplain, estuarine or marine) in developing countries. The community-based co-management process may also be relevant to small-scale fisheries in developed countries, as well as to other coastal resource (i.e. coral reefs, mangroves, seagrass, wetlands) management. Community-based co-management seems to be found in, and be most relevant to, developing countries due to the need for overall community and economic development and social empowerment, in addition to resource management (see Chapter 2, Section 2.8 for more discussion about different types of co-management).

1.2. What This Handbook Isn't

It should be noted that this handbook will not present a 'one size fits all' or blueprint process to co-management. It will also not present step-by-step procedures to co-management. This is not possible as there are as many approaches to co-management as there are communities. Each situation in which co-management may be used is different and will require a unique response. Rather, this handbook will present a process which is 'generic' in that it will provide the handbook user with a place to start and with an understanding of how 'a' process works and the variety of activities and methods than can be used to plan and implement co-management. This handbook will serve as a source of information for the practical planning and implementation of co-management at the community level. The process presented is meant to be 'a' process that should be adapted to the local situation. Each co-management activity and method will be covered in enough detail to allow the handbook user to understand and use it. However, it would be impossible to completely address in depth each activity and method in this publication. References and internet links will be provided for the user to obtain additional information.

1.3. Who This Handbook is For

This handbook is written for use by the initiators and facilitators of co-management – government, change agents, community leaders and community members, as well as motivated fishers. The handbook is written to allow each partner in co-management to clearly understand their role and responsibility in the co-management process, and how to relate to the other partners in co-management. The handbook is meant to be used to plan and implement co-management, as a reference source to specific co-management activities and as a training reference. More will be said about each partner and their roles and responsibilities in Chapter 4.

1.4. Using This Handbook

This handbook is meant to support a process of community-based co-management for fisheries. As discussed, it is not a step-by-step guide. Rather, the handbook provides ideas, methods, techniques, activities, checklists, examples, questions and indicators for planning and implementing community-based co-management.

The user should become familiar with the complete process presented, think about the situation where co-management will be considered for use, and adapt the co-management process to the community. The community may just be beginning co-management or it may already be implementing a process of co-management. Some or all of the activities and methods may be relevant. Other activities and methods may need to be undertaken. When you have identified an activity or method that is appropriate, you can read more about it in the handbook and in the additional references cited.

2 What is Community-based Co-management?

IDRC, R. Charbonneau.

This chapter contains a discussion of concepts and definitions about community-based co-management.

2.1. Co-management Defined

Cooperative management or co-management can be defined as a partnership arrangement in which the community of local resource users (fishers), government, other stakeholders (boat owners, fish traders, boat builders, business

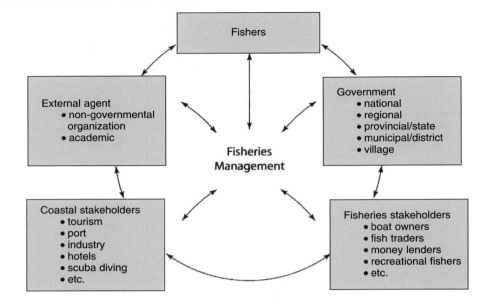

Fig. 2.1. Co-management is a partnership.

people, etc.) and external agents (non-governmental organizations (NGOs), academic and research institutions) share the responsibility and authority for the management of the fishery (Fig. 2.1). Through consultations and negotiations, the partners develop a formal agreement on their respective roles, responsibilities and rights in management, referred to as 'negotiated power'. Co-management is also called participatory, joint, stakeholder, multi-party or collaborative management.

Co-management covers various partnership arrangements and degrees of power sharing and integration of local (informal, traditional, customary) and centralized government management systems (Fig. 2.2). Fisheries co-management can be classified into five broad types according to the roles government and fishers play (Sen and Nielsen, 1996):

- *Instructive*: There is only minimal exchange of information between government and fishers. This type of co-management regime is only different from centralized management in the sense that the mechanisms exist for dialogue with users, but the process itself tends to be government informing fishers on the decisions they plan to make.
- *Consultative*: Mechanisms exist for government to consult with fishers but all decisions are taken by government.
- *Cooperative*: This type of co-management is where government and fishers cooperate together as equal partners in decision-making.
- *Advisory*: Fishers advise government of decisions to be taken and government endorses these decisions.
- *Informative*: Government has delegated authority to make decisions to

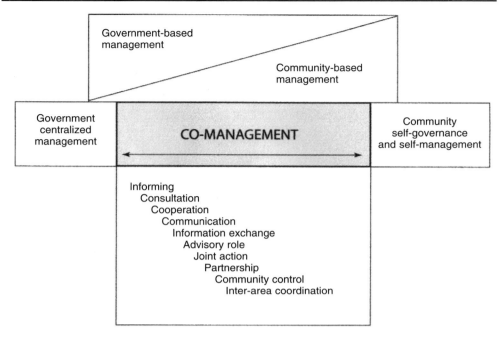

Fig. 2.2. Co-management integrates local and centralized government management systems.

fisher groups who are responsible for informing government of these decisions.

It is generally acknowledged that not all responsibility and authority should be vested at the community level (Box 2.1). The amount and types of responsibility and/or authority that the state level and the various community

Box 2.1. Community.

The term 'community' can have several meanings. Community can be defined geographically by political or resource boundaries or socially as a community of individuals with common interests. For example, the geographical community is usually a village political unit (the lowest governmental administrative unit); a social community may be a group of fishers using the same fishing gear or a fisher organization. A community is not necessarily a village, and a village is not necessarily a community. Care should also be taken not to assume that a community is a homogeneous unit, as there will often be different interests in a community, based on gender, class, ethnic and economic variations. Recently, the term 'virtual community' or 'community of interest' has been applied to non-geographically based communities of fishers. Similar to the 'social community', this is a group of fishers who, while they do not live in a single geographical community, use similar gear or target the same fish species or have a common interest in a particular fishery.

levels have will differ, and depend upon country and site-specific conditions. The substance of this sharing of responsibility and authority will be negotiated between community members and government and be within the boundaries of government policy. Determining what kind and how much responsibility and/or authority to allocate to the community level is ultimately a political decision in which government will always play a more decisive role. However, the key to co-management is negotiated power where the interaction of the state and non-state actors would be an important factor in defining a common and acceptable balance in sharing power and allocating responsibilities.

2.2. Co-management as a Process

There is no blueprint or model for co-management but rather a variety of arrangements from which to choose to suit a specific context. Co-management should be viewed not as a single strategy to solve all problems of fisheries management, but rather as a process of resource management, maturing, adjusting and adapting to changing conditions over time. A healthy co-management process will change over time in response to changes in the level of trust, credibility, legitimacy and success of the partners and the whole co-management arrangement. Co-management involves aspects of democratization, social empowerment, power sharing and decentralization. Co-management attempts to overcome the distrust, corruption, fragmentation and inefficiency of existing fisheries management arrangements through collaboration. Co-management is adaptive; that is, through a learning process, information is shared among partners, leading to continuous modifications and improvements in management. Through co-management, the partners actively contribute and work together on fisheries management. They share the costs and benefits and the successes and failures.

Co-management is not a regulatory technique, although regulations are used in co-management. It is a participatory management strategy that provides and maintains a forum or structure for action on empowerment, rule making, conflict management, power sharing, social learning, dialogue and communication, and development among the partners. Co-management is a consensus-driven process of recognizing different values, needs, concerns and interests involved in managing the resource. Partnerships, roles and responsibilities are pursued, strengthened and redefined at different times in the co-management process, depending on the needs and opportunities, the legal environment, the political support, capacities of partners and trust between partners. The co-management process may include formal and/or informal organizations of fishers and other stakeholders.

The establishment and operation of co-management can be complex, costly, time consuming and sometimes confusing. Research has shown that it may take 3 to 5 years to organize and initiate activities and interventions at the community level. It will also take this time for the partners to address concerns about legitimacy, trust, accountability and transparency.

Co-management can be considered as a middle course between the government's concern about social efficiency and equity and local concerns for active participation and self-regulation. Co-management involves a formal or informal agreement among partners to share power and to share the right to manage. Co-management can serve as a mechanism not only for fisheries management but also for community, economic and social development as it promotes fisher and community participation in solving problems and addressing needs. In some cases, co-management may be simply a formal recognition of a fisheries management system that already exists; some informal and customary community-based management strategies are already in place, operating side-by-side with formal state-level management strategies.

2.3. Stakeholder Involvement

Other than fishers, stakeholders (individuals, groups or organizations who are in one way or another interested, involved or affected (positively or negatively) by a particular action) that derive economic benefit from the resource (for example, boat owners, fish traders, business suppliers, police, politicians, consumers) should also be considered in co-management (see Chapter 4). These stakeholders often hold considerable political and/or economic influence in the community and over resource use and management. A proper balance of representation among stakeholders will prove crucial to the success of co-management (Box 2.2). A central question, however, is which stakeholders should

---

Box 2.2. Community and Stakeholders.

The term 'community' tends to abstract the diversity of interests among different groups of people. For example, some 'community consultations' are mainly attended by men but the results are considered as the output of the 'community'. The invisibility of women, especially in fisheries, is disregarded because fisheries are often viewed in a limited sense, i.e. mainly from the capture of marine or aquatic produce. While women rarely fish, they play important roles in pre- and post-production activities in fisheries.

The term 'stakeholders' could aggregate people too, with disregard for differential needs and interests. Stakeholders that only include the direct users of the coastal resources is a limiting conception. The non-recognition of women as direct users because they are not fishers (in the traditional sense) has resulted in their limited access, participation and benefits to coastal resource management projects and similar intervention focused on fisheries. The 1998 Philippine Fisheries Code has an initial recognition of women's interests in fisheries development and even it is limited in providing a single-seat for a women's organization in the fisheries council and still lacks an appreciation of women-specific issues in the agenda of the council. More work is definitely needed to better understand the contributions of women to fisheries.

Source: Tanyang (2001).

be represented and involved and how those representatives should be chosen. While it is useful to have representation of all stakeholders, a line must be drawn or the process will break down from the representation of too many interests. As will be discussed in Chapter 5, stakeholder analysis can help to identify those stakeholders who should be included in co-management. This question can be partially answered by determining the spatial scale at which co-management should operate. The best opportunity for co-management seems to occur at the local or 'community' scale (although national-level fisheries advisory bodies to government can also be effective).

2.4. Equity and Social Justice

Through co-management, equity and social justice in fisheries management is sought. Equity and the equitable sharing of power among and between government, fishers and other stakeholders in a community are often thorny issues in co-management. These issues are often visible along social and economic divides in a given community, such as gender roles. Those fishers who will receive the costs (and benefits) of management and regulation need to have a voice in decision-making about fisheries management. Equity and social justice is brought about through empowerment and active participation in the planning and implementation of fisheries co-management. Responsibility means fishers have a share in the decision-making process and bear the costs and benefits of those decisions. The theme of co-management is that self-involvement in the management of the resource will lead to a stronger commitment to comply with the management strategy and regulations. The mutuality of interests and the sharing of responsibility among and between partners will help to narrow the distance between resource managers and fishers, bringing about closer compatibility of the objectives of management.

2.5. Co-management and Common Property

Co-management is based on common property theory (Box 2.3). Co-management provides for the collective governance of common property resources. Common property regimes are forms of management grounded in a set of individually accepted rights and rules for the sustainable and interdependent use of collective goods. A collective good is defined as a resource that is managed and controlled by a group of users. A common property regime is composed of a recognized group of users, a well-defined resource that the group uses and manages, and a set of institutional arrangements for use of the resource. In some situations the group may formalize the institutional arrangements with an organizational structure for management. Common property represents private property for the group of co-owners (Gibbs and Bromley, 1987).

Common property resources share two key characteristics (Ostrom, 1991). First, these are resources for which exclusion (or control of access) of potential users is problematic. Second, the supply is limited; that is, consumption by one

---

Box 2.3. Property Rights Regimes.

The literature on property rights identifies four ideal analytical types of property rights regimes:

- State property: with sole government jurisdiction and centralized regulatory controls;
- Private property: with privatization of rights through the establishment of individual or company-held ownership;
- Communal property: in which the resource is controlled by an identifiable community of users or collectively, and regulations are made and enforced locally; and
- Open access: absence of property rights.

In reality, many marine and coastal resources are held under regimes that combine the characteristics of two or more of these types. The four property rights regimes are ideal, analytical types; they do not exist in the real world. Rather, resources tend to be held in overlapping combinations of these four regimes. Strictly speaking, pure communal property systems are always embedded in state property systems and state law, deriving their strength from them. Resource managers cannot function effectively unless they know the property rights regime they are dealing with.

Source: Bromley (1991).

---

user reduces its availability to others. It is also important to make a distinction between a resource unit and a resource system. A resource unit (what individuals use from the resource system such as fish) is not jointly used; while the resource system itself, the fishery, is subject to joint use.

Co-management is a governance arrangement located between pure state property and pure communal property regimes. It should be noted that while state law can enforce or strengthen communal property, it might not always do so. The level of help from the state will depend on its willingness to support communal property systems.

2.6. Institutional Arrangements and Collective Action

Common property regimes as collective resource management systems have been shown to develop when a group of individuals is highly dependent on a resource and when the availability of the resource is uncertain or limited (Runge, 1992). If the resource problem is repeatedly experienced, such as low or no catch, and if it exists within a single community of users, the fishers are likely to develop a collective institutional arrangement to deal with the problem. Institutional arrangements are sets of rights the fishers possess in relation to the resource and the rules that define what actions they can take in utilizing the resource. In the face of uncertainty in resource availability, fishers are more willing to group together to trade-off some benefit from individual use of the resource for the collective assurance that the resource will be used in a more equitable and sustainable manner (Gibbs and Bromley, 1987).

Institutions, through rights and rules, provide incentives for the group members to take certain actions to achieve a desired outcome. Institutional arrangements require an investment of time on the members' part to build. Coordination and information activities are initial aspects of building institutions. The transaction process of developing institutions will have costs. These transaction costs can be defined as the costs of: (i) obtaining information about the resource and what users are doing with it; (ii) reaching agreements with others in the group with respect to its use; and (iii) enforcing agreements that have been reached. For common property regimes, these costs are part of the collective decision-making process.

An individual member of the group relies on reciprocal behaviour from other members of the group regarding their adherence to the agreed-upon rules for management. An individual's choice of behaviour in a collective action (action taken by a group (either directly or on its behalf through an organization) in pursuit of members perceived shared interests) situation will depend upon how he or she weighs the benefits and costs of various alternatives and their likely outcomes. An individual's choices are often affected by limited information which leads to uncertainty and by the level of opportunistic behaviour (taking advantage of a situation in your own self-interest so as to get the benefit while bearing less of the cost) that individual resource users can expect from other resource users. Individuals also have differing discount rates (the value people put on the future benefits from the resource versus today); many poor fishers, for example, attribute less value to benefits that they expect to receive in the future, and more value to those expected in the present (Ostrom, 1991).

In some situations, individuals may have incentives to adopt opportunistic strategies to circumvent the rules and to obtain disproportionate benefits at the cost of others. Three types of opportunistic behaviour may occur: (i) free riding, (ii) corruption, and (iii) rent seeking. Free riders (holding back on one's contribution so as to get the benefit while bearing less of the cost) respond to incentives to engage in other activities while other members of the group work. Corruption can occur when incentives exist for rules to be changed for an individual through, for example, the provision of illegal payments. Rent seeking (the gaining of excess profits from the resource) can occur when an individual's assets, for example, property rights, increase in value through special advantages (Ostrom, 1992; Tang, 1992). The imperative of the common property regime is to establish institutional arrangements which reduce or minimize transaction costs and counteract opportunistic behaviour.

The principal problem faced by group members of a common property regime is how to organize themselves. That is, how to change from a situation of independent action to one of collective action and coordinated strategies to obtain greater joint benefits and reduce joint harm. A sense of 'commonality', commitment and compliance must be established for the collective good. There are two broad classes of problems that must be overcome by the collective group. The first, called appropriation problems, is concerned with how to allocate the resource units (i.e. fish) in an economically efficient and equitable manner among the resource users. The second, called provision problems, is focused on behavioural incentives to assigned duties to maintain and improve the resource over time.

Collective action does not occur where there is no organization that has authority to make decisions and to establish rules over the use of the resource. Note that institutions are not organizations. Organizations operationalize institutional arrangements (Bromley, 1991). There can be a variety of organizational forms for governing the resource, which may range from a government agency or enterprise to the fishers themselves. The common property regime will need to establish an agenda and goals which are to be achieved. This may include an identification of the problem or issue to be addressed, management and adjudication. The authority system to ensure that fishers' expectations are met is normally inherent in the organization.

Membership within common property regimes is not always equal. Some members may have fewer or lesser rights than others. Access to the resource, for example, may change or rotate for members through the year. Corresponding duties may or may not vary accordingly. The rights and duties of members of the group must be clearly specified.

Collective action entails problems of coordination that do not exist in other resource regimes, such as private property. In order to organize the harvesting, for example, fishers must develop rules to establish how rights are to be exercised. Rules give substance to rights, structure a situation, define the behaviour of the group's members and reduce conflict. Rules may create different incentive structures that affect cooperation or conflict among fishers (Tang, 1992). The type of rules that are devised will depend upon the severity of the problem the fishers face, the level of information they possess, sociocultural traditions, the extent of the bundle of rights they hold, the level of opportunistic behaviour, and the ease with which actions can be monitored and enforced. Rules require, permit or forbid some actions or outcome. Rules provide stability of expectations, and efforts to change rules can rapidly reduce their stability (Ostrom, 1991).

The institutions and rules fishers use may not always be the same as formal laws. The fishers may develop institutions and rules to meet their needs which are not legitimized by government.

For institutional arrangements to be maintained over time, it is important to develop workable procedures for monitoring the behaviour of fishers, enforcing against non-conforming behaviour with sanctions, and settling conflicts. The ease and costliness of monitoring rules devised to organize the fishing activity depend upon the physical nature of the resource, the rules-in-use and the level of conformance to the rules (Ostrom, 1990). The number of times that non-conformance must be measured affects the cost of monitoring. The ease and cost of monitoring will depend upon whether the fishers can monitor compliance themselves, as they fish or through self-monitoring incentives, or if they must establish more elaborate arrangements, such as external authorities.

Fishers who violate the rules need to have sanctions imposed upon them. What constitutes an effective sanction will vary depending upon the nature of the group of fishers. In most cases, sanctions are likely to increase with the severity of the offence (Ostrom, 1992).

Conflicts may arise within the common property regime and between users. The intensity and frequency of conflicts are likely to be closely related to the perceived relative scarcity of the resource. Several factors can lead to conflict, including: (i) absence of recognized rules, (ii) divergence in the interpretation of the rules, and (iii) outright trespass of a rule. Part of the institutional process must include a mechanism for discussing and resolving what is and is not a rule violation and how to settle the dispute. This can be done formally or informally. In general, for monitoring and sanctions to be effective, the fisher must have a stake in institutional processes and be involved in monitoring and enforcement (Townsend and Wilson, 1987).

Common property regimes and their associated institutional arrangements need to be dynamic in order to adjust to new opportunities, internal growth, externalities and institutional dissonance (Ostrom, 1992). Institution building is a long-term process and often is based on trial and error. Allocation rules, for example, may need to change as a result of poor compliance. The structuring of institutions must be an ongoing process to meet the changing conditions.

Whether or not local, self-governing institutions can be developed is often dependent upon governmental policies. In countries which do not recognize the rights of local community organizations or do not create opportunities for communities to organize themselves in a de facto manner, it is more difficult for fishers to successfully find solutions to collective action problems. Many governments are not willing to give up management authority over resources or do not believe that self-governing organizations can be successful. There is no single answer for how to resolve these differences.

2.7. Community-based Management

Community-based management (CBM) is a central element of co-management. There is some debate over the similarities and differences between co-management and CBM. Community-based resource management, as explained by Korten (1987), includes several elements: a group of people with common interests, mechanisms for effective and equitable management of conflict, community control and management of productive resources, local systems or mechanisms for capture and use of available resources, broadly distributed participation in control of resources within the community, and local accountability in management. Sajise (1995) has defined community-based resource management (CBRM) as 'a process by which the people themselves are given the opportunity and/or responsibility to manage their own resources; define their needs, goals, and aspirations; and to make decisions affecting their well-being.' Sajise (1995) further states that:

> CBRM as an approach emphasizes a community's capability, responsibility and accountability with regard to managing resources. It is inherently evolutionary, participatory and locale-specific and considers the technical, sociocultural, economic, political and environmental factors impinging upon the community.

CBRM is basically seen as community empowerment for resource productivity, sustainability and equity.

Ferrer and Nozawa (1997) state that:

community-based coastal resource management (CBCRM) is people-centered, community-oriented and resource-based. It starts from the basic premise that people have the innate capacity to understand and act on their own problems. It begins where the people are, i.e., what the people already know, and builds on this knowledge to develop further their knowledge and create a consciousness.

They further state that:

it strives for more active people's participation in the planning, implementation and evaluation of coastal resources management programmes. CBCRM allows each community to develop a management strategy which meets its own particular needs and conditions, thus enabling a greater degree of flexibility and modification. A central theme of CBCRM is empowerment, specifically the control over and ability to manage productive resources in the interest of one's own family and community. It invokes a basic principle of control and accountability which maintains that control over an action should rest with the people who bear its consequences.

Fellizar (1994) writes:

CBRM can be looked at in various ways. It can be a process, a strategy, an approach, a goal or a tool. It is a process through which the people themselves are given the opportunity and/or responsibility to manage their own resources; define their needs, goals and aspirations; and make decisions affecting their well-being. A strategy for achieving a people-centered development, CBRM has a decision-making focus in which the sustainable use of natural resources in a given area lies with the people in the local communities. CBRM is an approach through which communities are given the opportunity and responsibility to manage in a sustained way the community resources, define or identify the amount of resources and future needs, and their goals and aspirations, and make decisions affecting their common well-being as determined by technical, sociocultural, economic, political and environmental factors. It is a tool which facilitates the development of multilevel resource management skills vital to the realization of potentials of the community. Also, CBRM stands for people empowerment and achieving equity and sustainability in natural resource management. The key concepts are community, resources, management, access and control over resources, viable organizations and availability of suitable technology for resource management and utilization.

Rivera (1997) states that the CBCRM approach has several characteristics. It is consensus-driven and geared towards achieving a balance of interests. The emphasis is on communities and at its core is the community organization. It is a process of governance and political decision-making and it is geared towards the formation of partnerships and power-sharing. Rivera (1997) writes:

It can be argued that CBCRM is a politically negotiated process of making decisions on the ownership, control and overall policy directions of coastal resources. Questions of resource allocation, distribution of resource benefits and management arrangements among stakeholders will always have to be included. Moreover, CBCRMs central concern is the empowerment of groups and social actors and a sense of self-reliance at the micro-level that stimulates a more synergistic and dynamic linkage to the meso- and macro-levels. Further, it can be argued that CBCRM is the route to co-management. It is maintained that power issues are central to the formation of co-management schemes. Hence, partnerships between government and communities should take careful consideration of the capacities of communities in making and sustaining these partnerships.

Rivera states that in the Philippines, much of the work of non-governmental organizations (NGOs) on CBCRM can really be considered as co-management. Co-management is referred to by the NGOs as tripartite formation between government, community and the NGO. NGOs also refer to co-management as 'scaling-up', i.e. the recognition that the state cannot be ignored in sustaining local actions. The scaling-up efforts of NGOs include project replication, expansion of the geographic scale of management efforts (i.e. single community to multi-jurisdictional), building grassroots movements and influencing policy reform.

2.8. CBM and Co-management

The above definitions of community-based resource management show that while there are many similarities and differences between co-management and CBM, there are differences in the focus of each strategy. These differences centre on the level and timing of government participation in the process. CBM is people-centred and community-focused, while co-management focuses on these issues plus on a partnership arrangement between government and the local community of resource users. The process of resource management is organized differently too. Co-management has a broader scope and scale than CBM with a focus both inside and outside the community. The government may play a minor role in CBM; co-management, on the other hand, by definition includes a major and active government role.

Co-management often addresses issues beyond the community level, at regional and national levels, and of multiple stakeholders, and allows these issues, as they affect the community, to be brought more effectively into the domain of the community. CBM practitioners sometimes view government as an external player and adversary, to be brought into the process only at a late stage, if at all. This can lead to misunderstandings and lack of full support from government. Co-management strategies, on the other hand, involve government agencies, resource managers and elected officials early and equally, along with the community and stakeholders, developing trust between the participants.

When CBM is considered an integral part of co-management, it can be called community-based co-management. Community-based co-management includes the characteristics of both CBM and co-management; that is, it is people-centred, community-oriented, resource-based and partnership-based.

Thus, community-based co-management has the community as its focus, yet recognizes that to sustain such action, a horizontal (across the community) and vertical (with external to the community organizations and institutions such as government) link is necessary. Community-based co-management is most often found in developing countries due to their need for overall community and economic development and social empowerment, in addition to resource management.

One variation of community-based co-management is traditional or customary co-management. Such systems are or were used to manage coastal fisheries in various countries around the world. Existing examples in Asia and the Pacific have been documented over a wide discontinuous geographical range (Ruddle, 1994). Many of these systems play a valuable role in fisheries management and will be useful into the future, locally and nationally. Traditional or customary co-management is a formal recognition of the informal systems as used, for example, in Vanuatu and Fiji. Co-management can serve as a mechanism to legally recognize and protect these traditional and customary systems and to specify authority and responsibility between the community and government. It also involves a definition of shared powers and authority.

Stakeholder-centred co-management seems to be more common in developed countries, where the emphasis is to get the users participating in the resource management process. It can best be characterized as government–industry partnership that involves user groups in the making of resource management decisions. This category of co-management focuses on having fishers and other stakeholders represented through various organizational arrangements in management. Unlike community-based co-management, little or no attention is given to community development and social empowerment of fishers. Examples of stakeholder-centred co-management can be seen in several countries in northern Europe and North America (Nielsen and Vedsmand, 1995; McCay and Jentoft, 1996).

It should be noted that co-management and integrated coastal management (ICM) share many similarities such as the coordination of various stakeholders at different levels and an active role of government (Christie and White, 1997).

2.9. Advantages and Disadvantages

The potential advantages of co-management include:

1. A more transparent, accountable and autonomous management system.
2. A more democratic and participatory system.
3. More economical than centralized management systems; requiring less to be spent on management administration and enforcement, in the long run.
4. Through involvement in management, fishers take responsibility for a number of managerial functions.
5. Makes maximum use of indigenous knowledge and expertise to provide information on the resource base and to complement scientific information for management.

6. Improved stewardship of aquatic and coastal resources and management.
7. Management is accountable to local areas. Fishing communities are able to devise and administer management plans and regulatory measures that are more appropriate to local conditions. (Localized solutions to local problems.)
8. By giving the fishers a sense of ownership over the resource, co-management provides a powerful incentive for them to view the resource as a long-term asset rather than to discount its future returns.
9. Various interests and stakeholders are brought together to provide a more comprehensive understanding of the resource.
10. Since the community is involved in the formulation and implementation of co-management measures, a higher degree of acceptability, legitimacy and compliance to plans and regulations can be expected.
11. Community members can enforce standards of behaviour more effectively than bureaucracies can.
12. Increased communication and understanding among all concerned can minimize social conflict and maintain or improve social cohesion in the community.

Despite all these advantages, co-management has several disadvantages and problems, including:

1. It may not be suitable for every fishing community. Many communities may not be willing or able to take on the responsibility of co-management.
2. Leadership and appropriate local institutions, such as fisher organizations, may not exist within the community to initiate or sustain co-management efforts.
3. In the short-run, there are high initial investments of time, financial resources and human resources to establish co-management.
4. For many individuals and communities, the incentive(s) – economic, social and/or political – to engage in co-management may not be present.
5. The risks involved in changing fisheries management strategies may be too high for some communities and fishers.
6. The costs for individuals to participate in co-management strategies (time, money) may outweigh the expected benefits.
7. Sufficient political will may not exist to support co-management.
8. Unease of political leaders and government officials to share power.
9. The community may not have the capacity to be an effective and equitable governing institution.
10. Actions by user groups outside the immediate community may undermine or destroy the management activities undertaken by the community.
11. Particular local resource characteristics, such as fish migratory patterns, may make it difficult or impossible for the community to manage the resource.
12. The need to develop a consensus from a wide range of interests may lengthen the decision-making process and result in weaker, compromised measures.
13. There may be shifts in 'power bases' (political, economic, social) that are not in the best interests of all partners.

14. There are those who feel that co-management is too costly and time consuming and that other alternatives, such as stricter regulations, may be better.

15. There is always a possibility of unbalanced and inequitable sharing of power between government and communities and the use of co-management by some political leaders solely for their own purposes.

3 What is a Process for Community-based Co-management?

IDRC, M. Verde.

As stated earlier, the purpose of this handbook is to provide a practical reference on a process for community-based co-management for use by the initiators and facilitators of co-management. There is no blueprint or right process for community-based co-management. As already stated, there are a variety of different processes which can be tailored to meet a site-specific situation and context. In some cases there may already be a fisher organization capable of engaging in co-management, while in another there will be a need to organize fishers. In another case, an informal or customary fisheries management system may be functioning well in the community and the need

© International Development Research Centre 2006. *Fishery Co-management: A Practical Handbook* (R.S. Pomeroy and R. Rivera-Guieb)

is to reach a co-management agreement with government to provide support. In another, fishers may only now be discussing informally among themselves the need to engage in some form of collective action.

This handbook will present *a* process of community-based co-management, not *the* process of community-based co-management. The process to be presented in this handbook can be thought of as a generic process that can serve as a reference to those who want to initiate a community-based co-management programme. Start at the beginning or start in the middle depending upon the situation. The process presented is based on the authors' experience with community-based coastal resources management and co-management and on the experience of NGOs, development projects, governments and other institutions around the world. It takes lessons learned from successes and failures and provides a process to adopt, adapt or revise to the local situation and context. The activities presented in the process are present in some form in all the cases that the authors are familiar with.

Community-based co-management can be initiated in several ways. The process presented in this handbook is a community-level and community-initiated programme of community-based co-management that is implemented in partnership with an external agent (such as an NGO or academic or research institution) and government. Alternatively, it may be an externally initiated programme where the external agent or government agency identifies a problem (or problems), then establishes a programme in collaboration with the community. In another case, it may be initiated as part of a larger donor-assisted development programme, where there is little early consultation with the community. Co-management could also be initiated by the government fisheries department working with fishers to establish a national level fisheries advisory body to guide policy.

The ways in which the fishers approach the community-based co-management programme will depend, in part, on how the programme is initiated and their sense of ownership over the process. While the origin and initiation of the community-based co-management programme may differ (beginning or pre-implementation phase), it is the authors' experience that the implementation phase activities in the process presented in this handbook are usually present.

This chapter of the handbook will provide an overview of the phases, components and activities of the process. Much more detail of each will be provided later in the handbook.

3.1. A Process of Community-based Co-management

The implementation of community-based co-management can be viewed as having three phases: 'beginning' or pre-implementation, implementation and 'turnover' or post-implementation (Fig. 3.1).

It should be noted that many of the process activities described in this handbook are continuous and overlapping, especially during the implementation phase (Box 3.1). The co-management process is dynamic

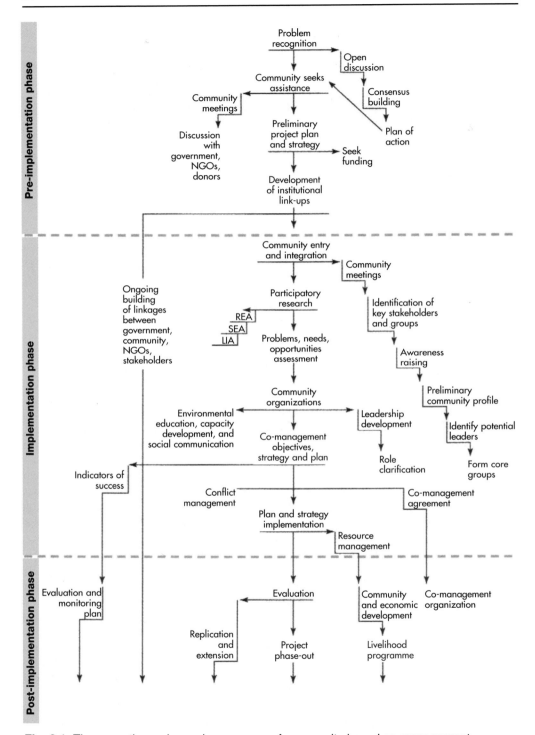

Fig. 3.1. There are three phases in a process of community-based co-management.

Box 3.1. What Interventions and Activities Should We Do?

As will be discussed, there are a large number of interventions and activities possible in the co-management process. It is often said by initiators and facilitators: 'We cannot do them all. We only have so much time and money. Which are most important to do?' It is extremely difficult to say which intervention or activity is more important than another. As has already been stated, each situation is different and will require a unique response. However, as a rule of thumb, it is suggested that as a minimum, community organizing and environmental education and capacity building should be undertaken. These are really the foundations of community-based co-management. They provide the basis for empowerment and participation.

---

rather than linear, often cyclic as it evolves, and adaptive. The three phases are not distinct but actually flow into and overlap each other. Some implementation activities may finish and enter a post-implementation phase while at the same time new implementation activities are being initiated.

3.1.1. 'Beginnings' or pre-implementation

The pre-implementation phase may start when fishers and other stakeholders recognize a resource(s) problem that may threaten their livelihood, income, and/or social and community structure. This is especially true where the resource users are highly dependent on a resource(s), availability of the resource(s) is uncertain or limited, and the users are highly identified with their fishing area. If the resource(s) problem, such as low or no catch, repeatedly occurs, if it exists within a single community of fishers and if the users are unable or unwilling to move to another fishing area, the fishers are more likely to take action to deal with the problem. Fishers will individually and collectively begin to discuss the problem and seek information, a process that often leads to consensus-building and agreement on a plan of action. This plan of action may be specifically designed to initiate a process of change and seek assistance to do so. The fishers may identify different stakeholders who should be included in this programme.

At this early stage, an enthusiastic individual(s) may step forward as the prime mover(s) of the co-management programme. The fishers may seek assistance from the government or external agents on possible solutions or courses of action to deal with the problem. These external agents and/or government may enter at this point to assist the community by organizing meetings and providing information to prepare a preliminary programme plan and strategy. A proposal for outside funding of the programme may be prepared. Initial approvals for the programme may be obtained from different levels of government and local officials and leaders.

At this point, linkages are established and strengthened between fishers, other stakeholders, external agents and government so that a partnership is developed. A formal or informal agreement for cooperation may be established at this time. The development and strengthening of these linkages and networking at institutional, group and individual levels, is a continuous process during the life of the co-management programme.

3.1.2. Implementation

The implementation of community-based co-management has four linked and complementary components:

1. Resource management.
2. Community and economic development/livelihoods.
3. Capacity building.
4. Institutional support/networking/advocacy.

Please note that no one component exists in isolation but they are linked and complementary to each other. For example, it is often found that solutions to fisheries management problems lie outside the fishery, thus requiring household livelihood and community development issues to be addressed.

The resource management component consists of activities to manage, protect, conserve, rehabilitate, regulate and enhance marine and coastal resources.

The community and economic development/livelihood component's purpose is to raise income, improve living standards and generate employment through alternative and supplemental livelihood development. It addresses both individual and family needs, as well as community needs such as community social services and infrastructure development, enterprise development, and regional economic development, which includes business and industrial development.

The capacity building component, aimed at individuals (both women and men) and organizations, involves empowerment and participation, education, training, leadership and organization development.

The institutional support/networking/advocacy component involves conflict management mechanisms, individual and organizational linkages, interactive learning, legal support, policy development, government restructuring, issue advocacy, networking with other communities and organizations, forums for sharing, and institution building and strengthening.

There are a number of specific activities and interventions in the implementation phase of community-based co-management.

Community entry and integration
Community entry and integration are usually the first steps in implementation. Field workers and community organizers (COs) (either from the community or from outside it) provided by the external agent begin to identify the main stakeholders, those groups and individuals with an interest in co-management, and facilitate the process. It is often difficult to determine who is and who is not a legitimate stakeholder and at what level in the co-management partnership they should be involved. The field workers and community

organizers establish initial relationships and credibility with community members, targeting project participants and local leaders at this time. The COs, working with the local people, identify and study the communication and participation structures in the community, including local social structures and power relations, forums for discussion and conflict management, communication barriers by gender and class, and participation in decision-making.

A series of meetings and discussions is held with resource users, stakeholders and government officials to share the concept and process of community-based co-management, to begin to develop communication and a consensus on their interests and concerns, and to build awareness about resource protection, management and rehabilitation. Community members actively participate in these activities. Other activities include identifying key individuals and groups to be involved in co-management, the organizing of a start-up team composed of community members (Box 3.2), answering questions about the programme, raising awareness about issues, the process and the programme, and participating in community activities such as fishing and local events. Government and local leader approvals are sought. At this

---

Box 3.2. Clarifying Groups and Organizations in the Co-management Programme.

There are four groups organized in the co-management programme. Two of the groups are disbanded or evolve into other groups as the co-management programme matures – core groups and co-management body; while the other two groups serve as the foundation of the programme – community organizations and co-management organization. More detail on each group is provided in the sections to follow, but in brief they can be described as:

- Core group – formed during the beginnings or pre-implementation phase. The core group is a small group of community members who early in the co-management programme act to facilitate the communication of information, develop dialogue with community members, facilitate community organizing, assist the community organizer in decision-making, and identify stakeholders. The core group is disbanded when the co-management organization is established.
- Community organization – organized by community members to represent their interests in the co-management programme. One or more community organizations may exist in a community depending upon its size, diversity and needs.
- Co-management body – established during the co-management planning stage, the co-management body is established to conduct and oversee the co-management plan process. It is composed of representatives of stakeholder groups. It may be disbanded at the end of the planning stage or evolve into the co-management organization.
- Co-management organization – once the co-management plan, strategy and agreement are finished, the co-management organization is established with the responsibility of managing the overall co-management programme. This is a permanent body that deals with conflict management, policy-making, monitoring, revising plans and agreements, financing and fund raising, information and data collection and analysis, education and research.

point, it is useful for the CO, working with community members, to conduct a feasibility analysis to determine whether a co-management arrangement would be possible. The legal, political, institutional, economic and sociocultural feasibility need to be considered (Borrini-Feyerabend, 1996). A household census may be conducted to collect socio-economic data on the community to initially identify problems, needs and opportunities. Community integration of field workers and community organizers can be a long process and requires those workers to have the skills, energy, commitment and personality to listen, share and work with the people of the community on an equal basis.

Research and participatory research

Participatory research is conducted to collect and analyse baseline data on the community, its people and its natural resources and to generate new knowledge. The baseline data are used in the preparation of development and management plans and strategies, for decision-making, for monitoring and evaluation and for process documentation. A participatory research process involves the people of the community, working with the researchers, in the design, collection, analysis and validation of the output. The participatory research process can also raise awareness and educate community members about their community and natural resources, as well as being useful in the formulation of potential solutions. Participatory research, which is conducted using a mix of scientific and rapid-appraisal methods, includes the collection of traditional and indigenous knowledge.

Participatory research can have four components, conducted in an interactive and iterative manner: (i) resource and ecological assessment (REA); (ii) socio-economic assessment (SEA); (iii) legal, policy and institutional assessment (LIA); and (iv) problem, needs, issues and opportunities assessment (PNIO). A REA provides a scientific and technical information base on the coastal and marine resources of the area. It usually includes three interrelated assessments: capture fisheries, coastal habitat (coral reef, mangrove, seagrass) and water quality. A SEA, which provides baseline information and a profile on social, demographic, gender, cultural and economic characteristics and conditions in the community, includes both stakeholder and conflict analyses. An LIA profiles the institutional arrangements (formal and informal rights and rules), organizational arrangements, legislation and policies and programmes (internal and external to the community) for coastal resources management. The PNIO is a participatory assessment of opportunities, problems, issues and needs, which is conducted by and with the stakeholders through a series of community meetings, key informant interviews, surveys and one-on-one discussions. Community members share with each other, as well as with government and external agents, ideas for their community's future and their vision on how to achieve that future.

Drawing on the baseline data from these four assessments, participants discuss the feasibility of developing a co-management agreement. The baseline data also serve as a basis for the future monitoring of the programme and for the evaluation of success and impacts.

Environmental education and capacity building

Environmental education and training, integral and ongoing activities of the community-based co-management programme, are the main methods of capacity building for community members and government. The external agent usually implements these activities, based on the assessments conducted earlier. The education and training should recognize and build upon the existing experience and knowledge of community members and government. Information is exchanged and the CO can learn from and with community members and government. Education methods, formal and informal, include small-group work, seminars, cross visits, role-playing, radio, video and fisher-to-fisher sharing of local knowledge. Environmental education is a priority goal of these activities, as is the building of community members' and government officials' and staffs' capabilities and confidence so they can make informed choices and decisions about problem articulation, management and development objectives, strategies and plans, and implementation.

Community organizing

Because community organizing is the foundation for mobilizing local human resources, community organizations and leaders are needed to take on the responsibility and authority for management and development activities. These organizations and leaders may already exist in the community, may emerge by themselves or may be newly established. Their focus is on participation, representation, and power sharing in the community. The members of any such group or organization must be willing to take on the responsibility. Existing organizations and leaders in the community are identified through the stakeholder analysis and LIA. Various types of organizational structures can become involved, including associations, cooperatives, unions, management councils and advisory committees, and may have differing levels of participation. Organizations may be formed at levels ranging from the fisher to the village to the municipal/district to the province/state. Organizing is often associated with work aimed at community members who are economically disadvantaged and/or with the least political power.

Education and training can empower the organization or leaders, developing their ability to take on management responsibility. Leadership development is an important part of this step, since strong and dedicated leadership is necessary if community-based co-management is to succeed. Existing community leaders, such as elected officials and senior fishers, play an important role but may be too closely tied to the existing community power structure to be advocates of improved equity. New leaders, often individuals with the motivation but not the means to take on leadership, can invigorate the process and increase its legitimacy. Terms of office for leaders should be short enough to decrease the possibility of corruption and power grabbing.

Adequate time must be provided for the organizing and leadership development processes. Lack of social preparation is often the cause of programme failure. It is during this step that the roles and responsibilities of organizations, leaders and stakeholders are delineated and clarified. Formal and informal fora for discussion and debate should be established, with stated place, time and rules for their meetings. Bridges are built between groups and organizations to improve communication and collaboration. The core groups and organizations advocate for support for policies, laws and local initiatives. Initial consultations and/or planning meetings are held among the partners to develop the co-management agreement.

Co-management plan and strategy

The community-level organizations, working in partnership with other stakeholders and the government, develop a resource management and community development plan whose objectives and strategies include a co-management agreement. Community members participate in the creation of the plan, validating its drafts along the way. The plan will include a common vision for the future, identification of a coordination mechanism and a financing strategy.

Reaching the co-management agreement may involve a series of meetings to negotiate and reach a consensus on its structure and to support management of conflicts. These meetings will involve identifying the key issues, as well as extensive bargaining and compromising in order to reach decisions.

The co-management agreement may include, specifically stated, a definition of roles, responsibilities and authority; identification of fora for meetings; conflict management mechanisms; and rule-making procedures. The agreement should be widely circulated to inform and obtain comments from relevant communities and stakeholders. A co-management body may be established at the end of the process of developing the agreement to represent all the partners. Participants would specify who is represented on the co-management body, what is its mandate and its level of authority and tasks. Indicators of success or monitoring and evaluation of the plan are specified. This can be done through a logical framework analysis (LFA) where outputs, activities, verifiable indicators and means of verification are stated.

The financial resources to implement the co-management plan should be identified early in the programme and made available before implementation. If external funding is needed to implement all or part of the plan, this is the time to identify a source and apply for the funds.

It should be noted again that the strengthening of linkages and partnerships and networking between resource users, stakeholders, government and the external agent is an ongoing and continuous process that extends beyond the implementation phase. The roles and responsibilities of the partners will change and adjust as the community-based co-management programme matures.

Institutional support will be sought, for example, to have formal recognition of the community organization or passage of a government ordinance legitimizing local institutional arrangements (rights and rules). The process of rights- and rules-making can be as difficult, yet as critical, as any other activity in the co-management programme.

Conflict management

Since conflicts will inevitably arise, the agreement must contain forms and mechanisms to address and resolve conflict. Conflict management is a process of dialogue and negotiation. A facilitator (a person who enables organizations to work more effectively), mediator (serves as a neutral party to assist stakeholders in finding a resolution to the conflict) or arbiter (makes a decision for the stakeholders at the request of the stakeholders) may be needed to guide the process towards constructive results. Participants should designate a forum for negotiation and agree on some rules for the process. They may generate and discuss various options for action, formally agreeing on one of those options. The conflict management mechanism should be multi-level to allow for an appeal process.

Plan implementation

The activities and interventions of the co-management plan are implemented through sub-projects. These may be resource management-related, such as marine reserve or sanctuary creation, mangrove reforestation, erosion management or fishing gear restriction. On the other hand, they may be about community development: such as a water well, a road or livelihood development, such as agriculture, aquaculture or small business enterprise. As needed, the responsibilities and rights of partners are clarified, conflicts are managed, and the agreement is enforced – possibly resulting in changes in the agreement or the development of a new agreement (Maine *et al.*, 1996).

Evaluation

Monitoring and evaluation should be central elements of the overall implementation process, although evaluation may also be conducted during the turnover or post-implementation phase. The indicators of success specified earlier are used in monitoring and evaluation, both done in a participatory mode. Participatory monitoring allows for adaptive management: interactive learning and a feedback system of success and failure while the programme is being implemented. It provides the community and external agents with information, during the life of the programme, so they can assess whether activities are progressing as planned, and whether modifications are needed. Participatory evaluation allows those internal and external to the community to evaluate programme objectives against results. It allows for planning for the future based on experience. The baseline information collected earlier in the programme can be used in the evaluation. The co-management agreement is also monitored on an ongoing basis, with the partners reviewing the results. Performance indicators may be used to measure progress of the co-management agreement, programme and implementation.

3.1.3. 'Turnover' or post-implementation

At this point, the programme, with assistance from an external agent and external funding, is fully taken over by the community and becomes self-sustaining. The post-implementation phase begins. The external agents work through a planned phase-out from the community and the other co-management partners. The phase-out should be planned and well understood by all to eliminate surprises and minimize problems. A self-sustaining funding strategy is put in place. New activities may be planned and implemented. Where feasible, people in other communities replicate and extend the results of the project. Fisher-to-fisher training and cross visits can be an effective way to train people in other communities. Project replication and extension can also enhance the credibility of the community-based co-management system in the eyes of the community and the co-management partners, since success often breeds success (White *et al.*, 1994; Johannes, 1998). Change and adaptation are central elements of post-implementation.

3.2. Doing It

In the next sections of this handbook, much more detail will be provided on the phases, components and activities described above.

4 Who are You and What is Your Role in Community-based Co-management?

R. Pomeroy.

4.1. Stakeholders

Stakeholders in community-based co-management can be defined as individuals, groups or organizations of people who are interested, involved or affected (positively or negatively) by marine and coastal resources use and management. This may originate from geographical proximity, historical association, dependence for livelihood, institutional mandate, economic interest, or a variety of other concerns. Stakeholders in coastal communities include fishers, their families and households, boat owners, fish traders, community-based groups, seasonal or part-time fishers, local business owners, local traditional authorities, elected government officials, representatives of government agencies, non-governmental organizations and others. There may be different stakeholders depending on their interests, their ways of perceiving problems and opportunities concerning marine and coastal resources, and different perceptions about and needs for management. Not all stakeholders have the same stake or level of interest in marine and coastal resources and thus may be less or more active and have different entitlements to a role in the co-management programme (see Chapter 6, Section 6.2.10 for more discussion on stakeholders).

Different types of stakeholders may be distinguished using some considerations and criteria, which include:

- Existing rights to marine and coastal resources;
- Continuity of relationship to resource (resident fisher versus migratory fisher);
- Unique knowledge and skills for the management of the resources at stake;
- Losses and damage incurred in the management process;
- Historical and cultural relations to the resources;
- Degree of economic and social reliance on the resources;
- Degree of effort and interest in management;
- Equity in the access to the resources and the distribution of benefits from their use;
- Compatibility of the interests and activities of the stakeholders;
- Present or potential impact of the activities of the stakeholders on the resource base (Borrini-Feyerabend, 1997).

Those who score high on several of these considerations and criteria may be considered 'primary' stakeholders and would assume a more active role in co-management, such as being on the management body. Secondary stakeholders may score on only one or two and be involved in a less important way, such as involvement on a consultative body (Borrini-Feyerabend, 1997).

Coastal communities are not all the same and are composed of people with different economic and social status, clans and family groups, language, ethnicity, customs and interests, which can create complexity for management. Even fishers from the same community and using the same fishing gear may have different interests. Coastal communities generally include a variety of stakeholders with divergent interests and views about co-management depending upon their involvement with the resources. These differences need to be recognized, understood and respected if co-management is to be promoted and involve the whole community.

4.2. Stakeholders in Community-based Co-management

For the purposes of this handbook, four key stakeholders or partners in community-based co-management are identified:

1. Resource users (fishers, family/household, community-based fisher group);
2. Government (national, regional, local);

3. Other stakeholders (community members, boat owners, fish traders, boat builders, business people, community-based groups, etc.);
4. Change agents (NGOs, academic and research institutions, development agencies, etc.).

This section of the handbook is written to discuss the role and responsibility of each of these four partners in the community-based co-management programme, and the relationship of the partners to each other.

4.2.1. Resource users

The local community is made up of individuals with differing interests in marine and coastal resource co-management. At the community-level, co-management projects usually have as their primary target fishers, that is, individuals who make their livelihood harvesting and using marine and coastal resources. The fishers are the individuals who, through their use of the resource, directly impact upon it and who are in turn directly impacted by management measures. Fishers are considered by many to be the real day-to-day managers of the resource, and as such, should be active participants in management. Fishers are usually the target of organizing and capacity-building activities.

Fishers' family and household are also stakeholders in co-management. Both the family and the household unit are identified, as, depending upon the culture, a household may include more than one family unit and several generations. In most fishing families, the fisher is usually a male. However, there are cases when decisions are made by both the husband and wife, and the spouse can be influential. Fisheries management programmes often leave out women, as the focus of the programme is the male fisher. Since women are involved in decision-making, and since women and children may be involved in aspects of production and marketing of the resource, it is clear that they are stakeholders, and should be partners, in co-management. Participation efforts, capacity-building efforts, and livelihood activities of the co-management programme, among others, should target women and, where appropriate, children.

Community-based fisher groups are formal and informal groups of local fishers established to support the social, cultural, economic and environmental interests of its members or of the community as a whole. These groups play an important role of bringing together fishers with similar interests concerning the resource and livelihood.

The main assets of community-based fisher groups are:

- Local knowledge, skills and resources;
- Built-in flexibility;
- Strong ties to the resource users, their families and the community;
- Direct responsiveness to local interests and conditions;
- Sociocultural cohesiveness;

- Capacity to serve the interests of its members;
- Confidence and trust of the local people (Borrini-Feyerabend, 1997).

Examples of community-based resource user groups are the people's organizations (POs) in the Philippines, the community fisheries (CF) organizations in Cambodia, and the mass organizations in Vietnam such as farmers/fishers union, women's union and youth union. In some cases, separate women's organizations are formed at the community level in the Philippines.

In general, the role of the community-based fisher groups in community-based co-management includes:

- Identification of issues and concerns of the community;
- Mobilization and leadership of co-management activities;
- Participation in research, data gathering and analysis;
- Participation in the planning, design and implementation of co-management activities;
- Community-based enforcement and self-regulation;
- Monitoring and evaluation;
- Advocacy to lobby for changes in or development of policy;
- Establish a people's movement for participation and change.

4.2.2. Government

Both the national and local government units (i.e. province/state, city, town, municipality, district, village) have jurisdiction over fisheries and coastal resources. Each government level has different mandates, authority and responsibility.

Increasingly, government policies and programmes stress the need for greater fisher participation and the development of local organizations to handle some aspect of resource management. Government must, however, not just call for more fisher involvement and participation, but also establish commensurate legal rights and authorities and devolve some of their powers. The delegation of authority and power sharing to manage the resource may be one of the most difficult tasks in establishing co-management. Government must not only foster conditions for resource user participation but sustain it.

As a first step, the national government must establish conditions for (or at least not impede) the co-management programme to originate and prosper. At a minimum, government must not challenge fishers' rights to hold meetings to discuss problems and solutions and to develop organizations and institutional arrangements (rights and rules) for management. Fishers must feel safe to openly hold meetings at their initiative and discuss problems and solutions in public fora. They must not feel threatened if they criticize existing government policies and management methods.

As a second step, fishers must be given access to government and government officials to express their concerns and ideas. Fishers should feel that government officials will listen and take action as necessary. As a third step, fishers should be given the right to develop their own organizations and to form networks and coalitions for cooperation and coordination. Too often there has been the formation of government-sponsored organizations which are officially recognized but ineffective since they do not represent the fishers. However, these may be the only type of organization a government may allow. Fishers must be free to develop organizations on their own initiative that meet their needs and that are legitimate to them. The issue is government's willingness to share authority and responsibility with the fisher organization and what function and form this will take.

One fundamental debate in co-management is the perception that fishers cannot always be entrusted to manage resources on their own. Unless government and officials who implement government policies can be convinced of the desire and the ability of fishers to manage themselves, not much progress can be made in co-management. The acknowledgement and acceptance of local-level management is partly the task of fishers to take on the new responsibilities, to organize themselves and, on the ability of the local community, to control the resources in question. On the other hand, communities and change agents often point out that government resource managers are reluctant to share authority. In this case, peers who have already 'bought into' co-management and/or a targeted education programme can be used to inform and, hopefully, persuade the resource managers to support co-management. While there are cases that show how politicians could use co-management to pursue their own personal goals and hang on to political power, it would be a mistake to interpret this for all government resource managers and officials. The success of co-management is fundamentally based on the trust extended among the partners and the commitment to collectively work towards a common vision.

A question for government is what management functions are best handled by the community, as opposed to the national or local government. Seven management functions have been identified that may be enhanced by the joint action of users and government resource managers at the local level: (i) data gathering; (ii) logistical decisions such as who can harvest and when; (iii) allocation decisions; (iv) protection of resources from environmental damage; (v) enforcement of regulations; (vi) enhancement of long-term planning; and (vii) more inclusive decision-making. No single formula exists to implement a co-management agreement to cover these functions. The answer depends on country-specific and site-specific conditions, and is ultimately a political decision.

The roles of the national government and national agencies in community-based co-management include:

- Provide enabling legislation to authorize and legitimize the right to organize and to make and enforce co-management;
- Determination of form and process and provision of decentralization;
- Recognition of legitimacy of community-based informal management systems;

- Address problems and issues beyond the scope of local co-management arrangements;
- Provide technical assistance;
- Provide financial assistance;
- Ensure accountability of co-management through overseeing local arrangements and dealing with abuses of local authority;
- Conflict management;
- Appeal mechanism;
- Backstopping local monitoring and enforcement mechanisms;
- Applying regulatory standards;
- Research;
- Training and education;
- Coordination role to maintain a forum for local co-management partners to interact;
- Gatekeeper in case the local co-management partners do not act upon their responsibility;
- Determination of allocation of management functions.

The main role of the local government unit is to support community-based co-management initiatives within its jurisdiction (Box 4.1). In many countries, local government units have a good deal of authority and responsibility to manage fisheries and coastal resources within their jurisdiction. This authority and responsibility may be historical or it may have been recently decentralized from the central government. There must be political willingness among the local political 'power elite' to support co-management. In addition, local government staff and officials must also endorse and actively support the co-management programme.

The roles the local government unit plays include:

- Supporting community involvement in community-based co-management;
- Approving local regulations and ordinances;
- Enforcement of regulations;
- Appeal mechanism;
- Providing technical assistance and staff;

Box 4.1. The Changing Role of Local Government in Co-management.

The role of local government (provincial/state, municipal, village) in co-management differs from country to country and the mandate provided to them from national government. In the Philippines, municipal and city government have an important role in coastal management because of the legal mandate to manage resources within municipal waters. In South Africa, the role of municipal government is very limited. Marine resources management is identified as a national competence. Aspects of fisheries management have been delegated to the provincial level, such as the provincial conservation authority in KwaZulu-Natal, but not to the municipal level, even though local authority could easily engage with provincial authorities. In South Africa, as in Mozambique and Angola, the national government is ultimately responsible for fisheries resource management.

- Providing financial assistance;
- Backstopping community-led functions, activities and mechanisms;
- Provide and/or support conflict management mechanisms;
- Ensure legitimacy and accountability of co-management;
- Engage in multisectoral and inter-local government unit collaboration;
- Facilitate and coordinate co-management planning and implementation;
- Provide a supporting environment for partner dialogue;
- Institutionalize co-management for fisheries and coastal resources in local waters (DENR *et al.*, 2001a).

4.2.3. Other stakeholders

A number of other members of the local community are directly and indirectly stakeholders in community-based co-management. These stakeholders will have varying interests in engaging in co-management, depending in part on their economic interests in the fishers and the resource. For example, a fish trader who is involved in a credit/marketing relationship with a fisher may not be supportive of the fishers organizing to be involved in co-management because it may be a potential threat to their relationship. It will be important to engage the fish trader in the co-management process through, at a minimum, education and discussion to understand the process, if not as an active partner. The role of each of these stakeholders in community-based co-management will be different and site-specific.

The stakeholders include:

- Business people: Local businesses such as boat owners, fish traders, fish processors, boat builders, hotel owners, recreational fishing guides, aquaculturists and shipping companies all use marine and coastal resources and have a strong economic stake in resource management issues. While they may not have to be direct participants, at a minimum they need to be consulted and educated so that they do not disrupt the co-management process. Local business people can provide an incentive and funding to resource users to manage resources.
- Community-based groups: Formal and informal groups of local community members established to support the social, cultural, gender, economic and environmental interests of its members or of the community as a whole. There can be a variety of community-based groups such as a women's group, religious organization, civic organization and service organization. There can also be community-based groups, other than community-based fisher groups, with a direct interest in the fishery and coastal resources, such as women's fish marketing association or coastal aquaculture organization. These community-based groups play an important role of bringing together individuals and groups with similar interests. These groups have local knowledge, skills and resources; strong ties to the community and the confidence and trust of the local people.
- Part-time and seasonal resource users: In many areas, there are a number of

part-time or seasonal resource users who depend on the resource for a part of their livelihood, income and food. These may be upland farmers who come to the coast to fish during the dry season or migratory fishers chasing small pelagics. While not a part of the resident community, they are a part of the community of resource users. Certain management measures can impact their livelihood, income and food security.

- Resource management organizations: In some countries, the law encourages the formation of collaborative resource management councils that have government and community representatives. These boards and councils provide technical expertise and serve in an advisory capacity to the government. These organizations can be important partners in co-management.

The roles of these other stakeholders include:

- Identification of issues and concerns of the community;
- Participation in planning and implementation;
- Providing incentives for certain behaviour;
- Dissemination of information;
- Fostering participation;
- Conflict management;
- Facilitation.

4.2.4. Change agents

Change agents include non-governmental organizations, academic institutions, research institutions, development agencies and similar organizations who act in a catalytic and facilitation role for community-based co-management. The change agent is considered to be a catalyst of change and to act as an intermediary between communities and external institutions, such as government, the general public and businesses. Change agents are meant to 'spark' endogenous change 'from within', not to carry out the change programme; this is the responsibility of the organized community (Rivera, 1997). The facilitation role is to empower and enhance the capabilities of the community to manage their lives and resources. Through the process of building community self-reliance, the change agent creates an impetus for community-based co-management.

Most change agents have a conservation and/or social development focus. They may be registered with the government and be officially recognized as a legal entity. The change agent should maintain relative objectivity and provide technical and analytical skills. The change agent provides a variety of services such as information and independent advice, ideas and expertise, education and training, community organizing, social development, research, advocacy, and finance and resource mobilization. Many change agents have staff, such as community organizers, who live and work in the community. A community presence can increase the level of trust between communities and the change agent and increase total participation.

In the beginning, the change agent may play a more dominant role with the community, but this should gradually diminish as the co-management programme progresses. It is important to note that change agents should realize that it is the people themselves who should be the driving force behind the co-management process; that genuine participation and change have to start with the people (Rivera, 1997). Problems can arise when people become too dependent upon the change agent or when the change agent directly interferes in the process, rather than guiding it or serving as a catalyst. Problems can also occur when the change agent's ideological views on development are not acceptable to the community or government. The role of the change agent in community-based co-management will differ by country, where, for example, an NGO is not allowed by the government to operate.

Change agents may choose to establish alliance with other change agents who have complementary skills, allowing them to implement more complex projects than they could by working alone. Alliances or networking increase the ability of change agents to learn from each other. It also allows the change agents to engage in advocacy to influence public policy.

Development agencies, as change agents, can provide funding and technical guidance for community-based co-management.

II Pre-implementation

5 'Beginnings' or Pre-implementation

'Beginnings' can be considered to be the pre-implementation phase of the community-based co-management programme. Beginnings of the co-management programme can be external to the community or internal to the community. Beginnings can have several activities including problem recognition and consensus, taking collective action, seeking information, community meetings and discussion, assessing the need and feasibility of co-management, preliminary plan and strategy, seeking funding and developing linkages (Table 5.1).

© International Development Research Centre 2006. *Fishery Co-management: A Practical Handbook* (R.S. Pomeroy and R. Rivera-Guieb)

Table 5.1. 'Beginnings' or pre-implementation.

Stakeholder	Role
Fishers	• Participate in discussions, meetings and dialogue • Communicate needs and problems • Provide inputs into initial agreement
Fisher leaders	• Seek information • Seek assistance • Lead preliminary plan and strategy • Organize meetings • Organize agreement • Develop linkages • Proposal and funding
Other stakeholders	• Participate in discussions, meetings and dialogue • Communicate needs and problems • Provide inputs into initial agreement
Government	• Provide information • Identify and provide assistance • Organize meetings • Organize agreement • Participate in meetings • Support community efforts • Develop linkages • Proposal and funding
External agent	• Provide information • Identify and provide assistance • Organize meetings • Organize agreement • Participate in meetings • Develop linkages and networking • Proposal and funding

5.1. External and Internal Beginnings

Community-based co-management may have a number of different types of beginnings, that is, the way in which a community-based co-management programme is initiated. The beginnings of co-management may be a response to conflict, an environmental crisis, new legislation, a conservation or development initiative, to take advantage of a funding opportunity, for political purposes, or to claim resource rights. The beginning of a co-management programme is often highly site- and context-specific and may involve a variety of different stakeholders. The actual beginning of a programme may originate from several different sources and combination of stakeholders and finally evolve into an action. The beginning phase of co-management may be time-consuming, difficult and costly. It is not possible to present all the possible scenarios of the beginnings of a co-management programme. In real life, all of the activities described below may not be undertaken or they may not be undertaken in the order described. For the sake of simplicity, two types of beginnings will be discussed. The co-management programme may be: (i) initiated external to the community, or (ii) initiated internally in the community. Several examples will be provided to help illustrate the range of possible beginning scenarios.

5.1.1. External beginnings

A community-based co-management programme, initiated externally to the community, is one in which the idea for the programme originates outside of the community. For example, the programme may begin where an external agent (i.e. NGO, academic or research institution) and/or government agency identifies a problem(s) (for example, poverty in fishing communities, overfishing, destructive fishing), plans to address the problem through co-management, further identifies an area or community in the country to focus the programme, then establishes the programme. In another case, the programme may be initiated as part of a larger donor-assisted development programme in the country in which community-based co-management is the intervention approach. Or it may be that the government has declared a protected area and wants to develop a co-management partnership to manage the area (Boxes 5.1, 5.2, 5.3 and 5.4).

Box 5.1. External Beginnings: Laughing Bird Caye and Friends of Nature, Belize.

Laughing Bird Caye is a sand and shingle island surrounded by a broad lagoon filled with a variety of coral reef structures located approximately 19 km from the town of Placencia. The area has traditionally been a fishing ground for local fishers. In the late 1970s, resorts in the area began to take tourists to the area. In the mid-1980s, local residents became concerned about declining resource conditions in the area, as well as talk about private development and an oil storage concession. In 1991, several community leaders organized the Friends of Laughing Bird Caye (FOLBC) to begin a consultation process in the community about the creation of a protected area and a national park. Through the efforts of FOLBC, the caye was declared a protected area in 1992. Working with the Department of Forestry, the FOLBC continued the consultation process to establish a management plan and buffer zone. FOLBC was registered as an NGO in 1996. In 1996, the caye was declared a national park and in 1998 a World Heritage Site. FOLBC continued its lobbying efforts for conservation of the area. In 2000, FOLBC signed a memorandum of understanding with the Department of Forestry to co-manage the national park. Under its co-management agreement, the renamed Friends of Nature (FON) assumed control of the regulations on zoning and the behaviour of users. FON is authorized to police within the management zones. FON appointed an advisory committee composed of people from the local villages in the area to assist in formulating policy on management. FON conducts environmental education programmes and maintains consultations with users of the national park and community members.

Source: Pomeroy and Goetze (2003).

Box 5.2. External Beginnings: Community-based Coastal Resource Management (CBCRM) Work in Pagaspas Bay, Batangas, Philippines.

A Filipino NGO called CERD started its CBCRM programme in Calatagan, Batangas province in 1992. CERD first asked the permission of *barangay* (village) officials to conduct a study in some portions of Pagapas Bay. Their staff stayed in the community for almost 8 months. Subsequently the research results were presented to community members in a bay-wide consultation. Farmers, fishers, teachers, students and local politicians attended the consultation. The various stakeholders agreed that illegal fishing was the main cause of poverty in Pagaspas Bay and they agreed to work together to solve this problem.

CERD staff trained some young people in the community and formed a drama group. These young people led other community members in requesting the local government to declare Pagaspas Bay as a marine reserve. Recognizing the strong commitment of the people to protect the bay, the local government eventually decided to declare Pagaspas Bay and the entire municipal waters of Calatagan as a 'marine reserve and in the state of rehabilitation'.

The next steps led to the formation of several fisher organizations along the bay. These steps were:

1. Selecting potential fisher leaders in the community that served as the core group;
2. The community organizers of CERD and the local leaders explained the benefits of organizing and invited interested people to attend a founding congress of a proposed fisherfolk organization;
3. The first municipal-wide congress of fishers was held and, subsequently, the Samahan ng Maliliit na Mangingisda sa Calatagan (SAMMACA) was formally organized;
4. Education and environmental awareness activities were held, e.g. basic course on marine ecological awareness and women's orientation; and
5. To expand organizing work and at the same time localize decision-making processes, *barangay* organizations of fishers were formed and became part of the municipal-wide SAMMACA.

Source: Aleroza *et al.* (2003).

In the case of a programme which begins externally to the community, there may or may not be early consultation and collaboration with the community in designing and preparing for the programme. In many externally initiated programmes, the details of the programme objectives, intervention approach (co-management) and the specific areas of the programme are decided in the design phase away from the community. For example, an NGO might specialize in co-management and have the resources to undertake a co-management programme in a community. The NGO would then conduct a community scoping activity (e.g. a needs assessment) to determine the feasibility of undertaking a co-management programme in the community. Once a programme community or area is identified, it is recommended that community members be consulted and allowed to participate as early in the programme life as possible in order to obtain their input, support and 'buy-in'. Although the project idea may not originate with the community,

Box 5.3. External Beginnings: The Sokhulu Subsistence Mussel-harvesting Project, South Africa.

The Sokhulu Tribal Authority lies on the north coast of KwaZulu-Natal between St Lucia and Richards Bay, immediately to the southwest of the Maphelane Nature Reserve, which forms part of the Greater St Lucia Wetland Park World Heritage Site. Prior to the promulgation of the Marine Living Resources Act in 1998, mussel harvest was controlled by a licence and bag-limit system and by specification of implement type. Traditional methods and quantities of mussel harvesting by subsistence gatherers were illegal under the legislation and were prevented by active law enforcement by the provincial conservation authority, the Natal Parks Board (NPB). Large-scale illegal harvesting of mussels by subsistence gatherers occurred at night along the coast, and conflict existed between subsistence gatherers and licensed recreational gatherers, and with the authorities. Violent clashes erupted between the Sokhulu community and gatherers and the NPB. The management staff felt that this situation could not persist and that the Sokhulu community should be approached in an attempt to try to find a solution. With outside funding, the NPD staff met with the local chief and an agreement was made to assemble all gatherers to discuss the matter. The meeting was well attended and an open and frank discussion was held on the problem. Despite some wariness on the part of the gatherers, they agreed to form a joint committee with the NPB staff to share information and generate an understanding between staff and the Sokhulu gatherers. The first few meetings were facilitated by an independent person, but once the initial mistrust and conflict was overcome, external facilitation was not necessary. Agreements were made to develop a sustainable harvesting system and to increase the capability of members of the fishing community to participate in management decisions. Decision-making within the subsistence zone is a joint endeavour, with the gatherers involved in decisions about the quota and in setting collecting rules.

Source: Harris *et al.* (2003).

the sooner they are aware that their community has been selected as a programme site, the sooner the community members can help shape specific programme objectives and strategies. If the programme objectives and strategies are kept relatively general at this early stage, community members can be given an opportunity to provide input into further programme design and planning and gain a sense of ownership of the project. Programme sustainability has been shown to improve when community members are given the opportunity to participate early in the programme design and planning stages and have an incentive to want to participate (Pollnac *et al.*, 2003).

Once a community is identified, an externally initiated programme may assist community members in problem identification, consensus building, accessing information and initial action planning. In some cases, an externally initiated programme, due to funding or donor demands, may immediately initiate implementation activities such as community organizer integration or education and capacity building (Box 5.5).

Box 5.4. External Beginnings: The Lake Malombe and Upper Shire River Fisheries, Malawi.

The Shire River and Lake Malombe are natural outlets of Lake Malawi. Biological studies suggested that fish stocks, such as the commercially important cichlid (*Oreochromis* spp., locally known as *chambo*) have been declining. This was attributed to the increase in use of narrow meshed seine nets which catch juvenile *chambo*. The livelihoods of fishers and their families were threatened by this decline in the fishery. Traditional management structures had largely died out with the commercialization of the fishery and government measures to regulate entry to the fishery and protect breeding and juvenile fish through closed seasons and legal mesh sizes proved ineffective. The Fisheries Department, with guidance and assistance from outside donor agencies, introduced the concept of co-management as a guiding principle of fisheries management under a pilot Participatory Fisheries Management Programme (PFMP). The strategy employed to implement the PFMP involved the creation of a Community Liaison Unit composed of fisheries extension staff and Beach Village Committees representing the fishing communities. The co-management strategy which was to be employed under the PFMP did not come from the fishing communities, and this caused some initial problems based on mistrust, struggles for power, and lack of a true partnership between the government and fishing communities. Although many of these issues have been resolved over time, the 'top-down' approach to initially implementing co-management was problematic.

Source: Hara *et al.* (2002).

Box 5.5. Selecting a Community with the Potential for Successful Implementation of Co-management.

With limited resources (financial, funds, time, people), initiators and facilitators of co-management often ask how they can select a community with greater chances for successful co-management. While there is never a 100% guarantee for success and sustainability of a co-management programme, there are a few characteristics about communities that can be considered before programme initiation. This is based on work by Ostrom (1990, 1992), Pomeroy *et al.* (2001), McConney *et al.* (2003b), Sverdrup-Jensen and Nielsen (1998) and Sowman *et al.* (2003). The characteristics include:

- Clearly defined boundaries;
- Group/community homogeneity;
- High dependency on the resource that is threatened/in crisis;
- Strong community ties to the sea and the resource;
- Individual incentive to participate due to livelihood being threatened;
- Existing organization to engage in co-management;
- Legal right to organize;
- Existence of decentralized authority;
- High level of indigenous knowledge about the resource;
- Supportive government and community leadership;
- Defined resource property rights;
- Resources are sedentary creatures that do not range far in their life cycle, distribution corresponds with human settlements, fall under one jurisdiction.

5.1.2. Internal beginnings

A community-based co-management programme initiated internal to the community begins from the fishers and other stakeholders in the community (Boxes 5.6 and 5.7). For example, the fishers, bearing different interests, concerns and capacities, come together to join forces and agree on a way to solve specific conflicts and problems and to claim specific rights. In contrast to the external beginning of a programme, this can be considered as a 'bottom-up' beginning for co-management. While ideally all stakeholders would be involved in deciding what to do, in reality a few stakeholders hold most of the authority and the means to set the co-management programme in motion. This type of beginning may be more promising in terms of sustainability of the programme as the fishers themselves have recognized a problem, they have an incentive to take action, and they take the lead to finding a solution. This is not to say that an internal beginning always leads to improved programme sustainability as compared to a programme with an external beginning. Additionally, to distinguish between an externally-initiated and an internally-initiated programme may be difficult because of highly complex interactions and interlocking relationships and interests between an external agent and the community. For example, a programme of the national government may want to expand its operation to a new community which has long recognized their problems but is not able to do anything about it. In this case, the idea to initiate some kind of management already exists in the community and this is put into action by an external institution with a similar idea and interest. The important point, whichever type or combination of beginning, is that the fishers and other stakeholders have an incentive (economic, social, political) to want to

---

Box 5.6. Internal Beginnings: San Salvador Island, Philippines.

San Salvador Island, with an area of 380 ha, forms part of the Masinloc municipality in the province of Zambales, on the western coast of Luzon, about 250 km north of Manila. White sand beaches surround the island, as do fringing coral reefs. In the late 1980s, when resource overexploitation, degradation and use conflicts reached a crisis point in San Salvador, some residents went in search of solutions to their problems. The highly centralized national government of the Philippines at that time was too distant to control the situation, while the San Salvador fishers themselves were too fragmented to embark on any collective action to avert resource degradation. Understanding that a serious problem existed that threatened their livelihood, several fishers and other community members requested the mayor to assist the village. External change agents were instrumental in helping to initiate new resource management measures. A Peace Corp volunteer who arrived in San Salvador in 1987 conceptualized the Marine Conservation Project for San Salvador, a community-based coastal resource management project for coral reef rehabilitation. In 1989, a local NGO led the project to establish a marine sanctuary. The NGO also assisted in environmental education, planning and leadership development.

Source: Katon *et al.* (1999).

Box 5.7. Internal Beginnings: *Hulbot-hulbot* Fishing in Palawan, Philippines.

Sometime in November 1995, three commercial fishing boats entered the municipal waters of *Sitio* Honda Bay in Puerto Princessa City, Palawan. The three boats used a gear locally known as *hulbot-hulbot* which involved dropping 1-tonne cement anchors on corals to scare the fish out of their hiding places and drive them into waiting nets. Even though the villagers knew the operations were illegal, they were either complacent or afraid of the operators. Before long, however, fish catch of the small local fishers began to drop dramatically. At first, only a handful of women from the area went to the NGO, Environmental Legal Assistance Center, Inc. (ELAC), to ask for some advice. They were on the verge of tears as they told their story of resource destruction. The consultation with ELAC led to action planning which included both legal and metalegal actions. Meetings with local and national government officials led to community action against the *hulbot-hulbot* fishers, and their eventual departure from the area. This initial action led the community to further organize into a Task Force Honda Bay which established a community monitoring and evaluation programme. Against this backdrop of activity, ELAC expanded its activities to include a broader based community-based coastal resource management programme for Honda Bay which included community organizing, policy advocacy, environmental education and paralegal training, socio-economic and marine resource profiling, land tenure issues and microenterprise development.

Source: Galit (2001).

participate and are given the opportunity to take ownership of the co-management process early in the project life.

5.2. Problem Recognition and Consensus

The beginning phase of a community-based co-management programme usually starts when fishers and other stakeholders are concerned about or recognize a resource(s) problem and/or conflict that may threaten their livelihood, income and/or social and community structure. This is especially true where the fishers are highly dependent on a resource(s), availability of the resource(s) is uncertain or limited and the users are highly identified with their fishing area (Runge, 1992). If the resource(s) problem, such as decreasing or no fish catch, repeatedly occurs over a period of time, if it exists within a single community of fishers and if the users are unable or unwilling to move to another fishing area, the fishers are more likely to take action to deal with the problem (Gibbs and Bromley, 1987; Ostrom, 1992). At this point, fishers may individually and collectively begin to discuss the problem informally, either while fishing, at the landing area, or while relaxing after fishing. Different points of view, interests and concerns about the problem may arise in the discussions. Fishers may also discuss the problem at more formal community or government meetings or with respected community members or government officials. After a period of discussion, the fishers may come to a consensus that there is a problem that needs solving and informally (or formally) organize themselves to take action.

5.3. Taking Action

With an informal consensus among the fishers and other stakeholders about the problem, an informal (or formal) agreement is reached that action needs to be taken. At this early stage, the development of the co-management programme is often based more on informal than formal relationships among people. The fishers may initially organize themselves and assign individual tasks to begin developing a plan and strategy for action. At this early stage, an enthusiastic individual(s) may step forward to lead the process. This individual is often visionary, dedicated and enthusiastic, and may have worked for a long time to prepare for co-management. A senior fisher, representative of a community organization, community leader or government official may be asked to assist. The leaders are usually self-selected based on strong personal motivation. These individuals may later become part of the core group which guides the whole co-management programme (see Chapter 6 and Box 3.2). The initial leaders will need to be willing and able to dedicate energy, time and creativity to the co-management programme.

This initial action may involve seeking information and/or assistance on what to do next or may involve putting together a more formal plan and strategy. Information may be sought from people in other communities about whether they have a similar problem and what they have done about it. Or information and/or assistance may be sought from the government (local, provincial/state and/or national) and/or an external agent (NGO, academic or research institution, religious institution) on possible solutions or courses of action to deal with the problem. At this stage, the group of fishers may discuss and identify other stakeholders who should be included in this process. In some cases, the plan of action may be limited to just the fishers and specifically designed to solve a single problem or issue, leaving the government and other stakeholders initially outside of the programme.

It should be recognized that co-management is a political process in which different stakeholders in the process may have different aims ranging from more equity in management to co-opting the rights of others. Power differences among stakeholders must be understood and recognized in order for the process to develop.

5.4. Information

Information is important at this early stage. The fishers are seeking information on who can help them, what assistance different institutions and organizations can provide, what has been done elsewhere, what has worked and what has not worked and where funding support can be obtained. This information may be available from individual people or institutions, or in publications or other forms of media (Box 5.8). This is an especially critical time in the programme as the fishers may not know exactly what they want or need or who can best assist them. They may have to talk to a number of people and institutions before they identify someone who can provide them assistance. It is often easy for the fishers to get discouraged if it becomes difficult to find assistance. The government may not be capable at this time of providing the assistance that the fishers need. NGOs with experience on fisheries and coastal issues may not be operating in the area. The fishers will need to be creative and diligent in their search for assistance. For example, an NGO which deals with health issues may be operating in the community and, while they do not deal with fisheries issues, they may be able to network with other NGOs to identify an NGO capable of assisting the fishers. Or government may have an extension agent who is willing to learn and assist. Or a lecturer at the local college or university may be interested and willing to assist.

Box 5.8. Sources of Information and Assistance on Fisheries and Coastal Resources and Management.

- National fisheries agency
- Local extension agent
- University
- High school
- NGO
- Mayor
- National environmental agency
- Doctor

5.5. Community Meetings and Discussion

The fishers may be organized enough to utilize the information obtained to prepare a preliminary project plan and strategy. However, in many cases, an external agent and/or government official may enter the community at this point to begin to provide assistance. Outside assistance is often needed as the fishers may lack the capability to act collectively. However, in situations where fishers are motivated by a crisis, for example, they can act collectively, although the effort may not be sustainable. The external agent and/or government official may organize community meetings and one-on-one and small group discussions to introduce themselves, provide information and discuss the local situation (Box 5.9). The meetings and discussions are meant to begin to gather and organize information on the local situation and fishers attitudes and perceptions regarding the resource and management. They are also meant to begin to inform and educate the fishers about the resource and possible solutions to problems, such as co-management. They are also meant to open and maintain communication among the various stakeholders. These meetings and discussions may initially be rather informal and unstructured as

> **Box 5.9.** Community Meetings and Small Group Discussions.
>
> Community meetings and small group discussions should be open and allow the participants to voice their views and ideas. It is especially important in this early stage of co-management that participants have a free exchange of ideas and feel that someone is listening to them. The meeting and discussion should be for the participants, with the external agent and/or government facilitating it for the participants. There should be an exchange of information from the external agent and/or government and they should share their insights and information to increase the level of awareness of the participants and on their role as co-managers.

everyone gets to know each other and to establish rapport and to determine if they can and want to work with each other. There may be one or several meetings and discussions depending upon the needs of the community. If not satisfied or confident in the external agent or government official, the fishers may look for other sources of assistance.

5.6. Assessing the Need, Feasibility and Suitability of Co-management

At this stage, the idea of co-management may be introduced and discussed. It is important to assess the need for, feasibility of and suitability of co-management for the community. The decision to engage in the co-management programme is both technical and political, and should be based on an analysis of technical and political needs (Borrini-Feyerabend *et al.*, 2004).

It is advisable to pursue co-management when:

- The active collaboration of the partners is essential for the management of the fishery and coastal resources;
- Access to the fishery and coastal resources is essential for livelihood, food security and cultural survival.

It is appropriate to pursue co-management when:

- The local stakeholders have historically enjoyed customary/legal rights over the area at stake;
- Local interests are strongly affected by the way in which the fishery and coastal resources are managed;
- The decisions to be taken are complex and highly controversial;
- Previous management approaches have clearly failed to produce the expected results;
- The various stakeholders are ready to collaborate and request to do so;
- There is ample time to negotiate.

It may be inappropriate to pursue co-management when rapid decisions are needed as in a crisis or emergency situation.

A judgment will need to be made whether or not the potential benefits of co-management outweigh the costs. If they do, a more detailed feasibility analysis may follow to determine that the conditions for co-management are in place (Borrini-Feyerabend, 1996).

When a group of individuals has determined that co-management is needed and desirable, they may wonder whether it is feasible in the particular context. In this case, it may be expedient to ask the following questions (Borrini-Feyerabend *et al.*, 2000, pp.17–18):

- Is co-management legally feasible? Who has the mandate to control the resources? Can a co-management approach be accommodated with the existing customary/legal framework? Examine traditional and modern laws, regulations, permits, etc.
- Is co-management politically feasible? What is the history of resource management and use in the area? Examine current political will and stability, the capacity to enforce decisions, the confidence in the participatory process, the presence of phenomena such as corruption and intimidation, etc.
- Is co-management institutionally feasible? Is there a chance of building a co-management institution in the area? Examine inter-institutional relations and their possible conflicts, existing examples of multi-party resource management organizations and rules, the capacity of stakeholders to organize themselves and express their choice of representatives to convey their interests and concerns, etc.
- Is co-management economically feasible? Are there economic opportunities and alternatives to the direct exploitation of natural resources? Examine local opportunities to reconcile the conservation of nature with the satisfaction of economic needs, examine the extent of poverty in the region, the availability of capital for local investment, etc.
- Is co-management socioculturally feasible? Are or were there traditional systems of natural resource management in the context at stake? What are (or were) their main features and strengths? Are those still valid today? Are the traditional natural resource management systems still in use? Regardless of whether the answer is yes or no, why? Who is keeping them alive? What is specifically sustaining or demeaning them? If they are not being used any more, does anyone have a living memory of the systems (for instance, are there elders who practised them and still remember clearly 'how it was done')?
- Examine the current population status, population dynamics and structure, the main sociocultural changes under way.
- Examine social and cultural diversity amongst the institutional actors and the history of group relations among them.
- Examine factors affecting opportunities for social communication, including:
  - Language diversity;
  - Varying degrees of access to information;
  - Different attitudes, for example, with regard to speaking in public or defending personal advantages;
  - Traditional and modern media currently used in the particular context.

Feasibility conditions do not need to be absolutely ideal to decide to embark on a co-management process, but thinking about feasibility factors gives a good idea of the obstacles and issues to expect along the way.

Certain questions can be asked in order to determine whether the proper institutional structure and support systems can be set up to make co-management a viable option for the community.

- Is the unit (the issue, the resource, the geographic area) of co-management definable? Does it relate to stakeholder interests? Is the scale manageable?
- Can participants agree on a set of objectives for the co-management regime?
- Can criteria for membership as co-managers and stakeholders be established? At least one member of the co-management regime must be a government representing the public interest.
- Is there a legally mandated basis for the co-management regime, or can a mandate be created?
- Can the co-management regime be financially supported, and does it include financial or in-kind contributions from members?
- Can members agree on a process for consensus-driven decision-making to address administrative matters, such as information sharing, capacity building, public communication, dispute resolution, evaluation, and for revising the process to reduce the risk of unsustainability?
- Will the parties volunteer, recognizing that there is mutual value to be gained from the co-management system? (National Roundtable on the Environment and the Economy, 1998).

It should be noted that if a full-scale co-management is not relevant or feasible at the time, at a minimum it would be useful to begin to organize fishers so that they could engage in future programmes.

5.7. Preliminary Plan and Strategy

Out of these meetings may come a consensus for the development of a preliminary programme plan and strategy. The initial group of leaders will direct this planning. This preliminary plan and strategy will identify the next steps to be taken in setting up and implementing the community-based co-management programme, the partners and stakeholders, and the partners' initial roles and responsibilities. This preliminary plan and strategy will differ from the co-management plan and strategy to be developed later in the process and which will include more specific management and development activities.

An initial agreement for cooperation between the group of fishers and the community and the external agent or government is prepared. This agreement specifies the services and assistance to be provided and the role and responsibility of the partners. It encourages information sharing, coordination and participation in planning and implementation. The preliminary plan will also assess the human and financial resources currently available and needed to undertake co-management. Are there resources to undertake all activities, including the community entry and integration phase? Thought needs to be given to the human and financial resources needed to undertake co-management.

5.8. Seeking Funding

If necessary, a proposal for funding of the programme may be prepared (Box 5.10). While a co-management programme can proceed without outside funding, such as with resources (people and funds) available in the community, most often some additional funding is needed to carry out the programme (Box 5.11).

The initial group of leaders may request assistance from government or an external agent in preparing the proposal. The human and financial resources needed to prepare and implement the co-management programme are identified. The preliminary plan and strategy will provide information on the objectives and approaches to prepare the proposal and to estimate resource needs. Several potential funding sources will need to be identified and contacted. It may be necessary to contact other programmes or external agents with experience with co-management to provide information to estimate human and financial resource needs. Depending upon how organized the fishers are, they may lead this funding search, but in most cases it will be led by the external agent or government. In most cases, a legal entity, such as a

---

Box 5.10. Proposal Writing.

Successful grant writing involves the coordination of several activities, including planning/preparation, identifying a funder, searching for data and resources, writing and packaging a proposal, submitting a proposal to a funder, and follow-up with the funder. The proposal should include project goals and identify the specific objectives that define how the goals will be accomplished.

The proposal should include the following sections:

- Project title/cover page;
- Summary/overview;
- Background information/problem statement;
- Goals and objectives;
- Target audience;
- Methods/design/strategy;
- Staff/administration;
- Impacts/benefits;
- Available resources;
- Needed resources;
- Evaluation plan;
- Timeline/workplan;
- Budget;
- Appendices.

Box 5.11. Potential Funding Sources.

A number of potential funding sources exist at the local, national and international levels to support a co-management programme. These include:

- Government – local, state/provincial, national;
- Private philanthropic foundations;
- Individuals;
- Companies;
- International development agencies;
- Non-governmental organizations;
- Development banks;
- Coastal-dependent businesses (e.g. beach resorts, boat charter companies, dive companies).

---

community organization or NGO registered with government, will be needed in order to submit the proposal and for the donor to provide funding.

5.9. Approvals

Initial approvals for the programme may need to be obtained from different levels of government and local community officials and leaders to move the programme forward and to submit a proposal.

5.10. Linkages

At this point, linkages are established and strengthened between and among fishers, other stakeholders, external agents and government so that a partnership is developed. Building partnerships is important to mobilize resources and energy of various stakeholders to achieve a common vision and goal. These are institutional and personal linkages meant to enhance the co-management programme. The development and strengthening of these linkages, at institutional, political and individual levels, is a continuous process during the life of the co-management programme. These linkages are important because they begin to give each partner a sense of the abilities and credibility of the other partners. Developing trust, credibility and friendship will often take time as the partners may not have worked directly with each other in the past or they may have reason for suspicion.

5.11. Moving to Implementation

This is now an in-between time before implementation. If resources such as money and people are available, initial implementation activities may begin.

If resources are not available, implementation will need to wait until proposals are funded or other arrangements are made to obtain the necessary resources. This can be a frustrating time as the partners are ready to go but cannot yet activate their plans and strategies. At a minimum, regular communication must be maintained between the partners. The momentum of early interest and action must be maintained. Meetings can be held to update each other on the status of funding proposals and other issues. Partners may obtain and share new information on the resource, the community, possible solutions, co-management, or other topics of interest and concern.

The external agent or government may begin to assess the human and financial resources available for co-management and their own capabilities and needs to engage as a partner in co-management. The fishers may want to begin organizing themselves (if they are not currently organized) to participate in co-management.

III Implementation

Implementing community-based co-management is always context-specific. For example, one community may have an existing and well-functioning fisher organization while another needs assistance in organizing. Another community may have a specifically defined set of objectives while another needs to develop a plan and strategy. During the 'beginnings' or pre-implementation phase, a preliminary plan and strategy is prepared to guide the implementation phase in the community. The goals and objectives identified in the preliminary implementation plan and strategy will be reflected in the types of implementation activities which will be undertaken and in the timeline for implementation. However, it is important to be flexible and adaptive as needs, issues and opportunities will change.

The implementation activities to be discussed below are present in some form in most community-based co-management programmes. It should be noted that the implementation of community-based co-management is not a linear process but involves a number of activities that may or may not occur sequentially or concurrently. Every community has its own unique situation and context and this should determine the activities in the co-management implementation process for that particular community. The implementation activities presented below should be adapted to the community.

The implementation phase generally begins when resources are available and the partners are ready. Implementation activities may be started with the resources (money, time, personnel) currently available. In other cases, implementation may not begin until adequate funding is obtained.

The general types of implementation activities include: (i) community entry and integration (Chapter 6); (ii) research and participatory research (Chapter 7); (iii) environmental education and capacity development (Chapter 8); (iv) community organizing (Chapter 9); (v) co-management plan and strategy (Chapter 10); (vi) conflict management (Chapter 11); and (vii) co-management plan implementation, including evaluation (Chapter 12). Each of these implementation activities will be discussed in the following chapters.

6 Community Entry and Integration

J. Parks.

Community entry and integration is normally led by the external agent. Community entry and integration establishes the initial working relationship between the community and the external agent and/or government involved with the programme (the implementation phase may be led by an external agent and/or government; for ease of comprehension, only the term external agent will be used in this publication) (Table 6.1). The external agent may introduce a community organizer or extension worker to the programme (Box 6.1).

Table 6.1. Community entry and integration.

Stakeholder	Role
Resource users/community	• Attending meetings and briefings • Prepare workplan
Local government	• Attend courtesy calls • Participate in meetings and discussions • Assist in organizing community meetings • Assist in identification of community boundaries • Assist preparation of workplans
Other stakeholders	• Attend meetings and briefings • Provide support
Change agent/community organizer	• Organize courtesy calls to government leaders • Orient to situation • Organize community meetings • Observation • Identify stakeholders • Prepare workplan

Community entry and integration entails a number of activities to initiate the project including:

- Formally introducing the programme to the community;
- Answering questions about the programme;
- Establishing rapport with the community;
- Participating in community life;
- Identifying roles of programme partners;
- Core group formation;
- Organizing and attending meetings, training and awareness-raising sessions;

Box 6.1. Community Entry and Integration in Vietnam.

The Center for Rural Progress and the International Marinelife Alliance-Vietnam, Vietnamese NGOs, initiated community-based coastal resource management projects in two villages in Vietnam. In assessing the lessons learned, it was reported that when the project initiator is coming from outside the community, the key is the ability to build a strong relationship directly with the community and local authorities. This allows for a certain level of legitimacy when working at the village level and sets the groundwork for a strong partnership. A major part of the project is the training and hiring of local authorities and members of community organizations to work on the local assessment. The 'newly-trained' people can facilitate the field work and make positive contributions which enhance the overall project. A further spin-off is that by training and working directly with these people, their own capacity is strengthened.

Source: Center for Rural Progress and the International Marinelife Alliance-Vietnam (2003).

- Collection of baseline data on the community;
- Stakeholder identification;
- Meeting with local leaders and government officials;
- Obtaining government approvals; and
- Initiating the programme with the community.

As with other activities of the co-management programme, the timeframe for community entry and integration will depend on the level of effort needed and objectives and outcomes set by the community, the programme and the community organizer (CO).

6.1. The Community Organizer

The community organizer (CO) is usually a staff member of the external agent or a government staff person, such as from the fisheries or extension department (Box 6.2). The CO may come from within or outside the community. The CO is a facilitator (a person who enables organization to work more effectively) for the co-management programme. In many cases, the CO will work and/or live in the community for many months or years to plan and implement co-management. The CO will continue to work with the community until the CO and the community feel that external assistance on a daily basis is no longer needed for sustainability of the programme. The CO should facilitate rather than impose.

Box 6.2. No Community Organizer.

In a number of cases, there may be no CO to facilitate the co-management programme. There may only be a willing government fishery officer or college teacher to facilitate the programme. Co-management can still move forward. Most important, the individual will need to be open-minded, creative and respectful. The fishery officer or teacher can initially serve to mobilize and energize the co-management programme. This individual will need to learn new skills, such as community organizing and participatory research, in order to support the co-management programme. This can be achieved by seeking out assistance and information from NGOs or projects functioning in the area or country. The development of a network with other individuals or organizations working on the same issues can help in the learning process.

The external agent and CO will play a central role in facilitating the co-management programme due to their knowledge and experience in community organizing, participatory methods, mobilization, education and information dissemination, and planning. The external agent and CO should have a phase-out strategy from the start of their involvement with the programme as the goal of their effort is to empower the community to manage the co-management programme and their resources themselves. The community should not become dependent upon the external agent to lead the co-management programme. As stated earlier, the role of the external agent should be to facilitate the co-management programme. The external agent and CO may need to continue to visit the community and provide assistance as needed after the phase-out (Boxes 6.3, 6.4, 6.5 and 6.6).

Box 6.3. Case Study: The Local Community Organizer Model in Bolinao, Pangasinan, Philippines.

The Local Community Organizers (LCO) model was implemented in Bolinao, Pangasinan, Philippines. It shows that while a CBCRM project may be externally initiated, promoting self-reliance and development of local capacity to prepare local organizations for the eventual phase out of the NGO or a project is important. With the LCO model, fisherfolk leaders were nominated by their organizations as 'community scholars' who learned the skills and attitudes of a community organizer. They were regarded as representatives of the PO, not the NGO; they regarded organizing as a learning opportunity that was awarded to them as a public trust. The LCO development emphasized sharing, confidence-building and learning-by-doing activities. The advantages of the LCO programme come from three aspects: local organizers are given opportunities to make unique contributions to organizing work, it is replicable and it is sustainable. With the LCO model, training and organizing became continuous and integrated in the learning process. Local leaders were movers and doers in organizing their own communities. Expanding organizing work was more cost-effective too because the LCOs organized adjacent communities by themselves. When Haribon Foundation, the facilitating NGO in Bolinao, eventually phased out, the PO federation continued to work on their own even up to now. For KAISAKA, the PO federation, their work continues even with the termination of the CBCRM programme.

Source: Arciaga *et al.* (2002).

Box 6.4. Local Community Organizers, Philippines.

In the Philippines, some NGOs select members from the community to act as local community organizers or LCOs. The tasks of the LCOs are essentially the same as the external COs, although they would usually focus on explaining the project objectives and activities to different community members thereby increasing awareness among the people. They also play a critical role in continuing the task of meeting people in the community, even informally, while the CO is away. In Bolinao, Pangasinan, Philippines, the LCO was distinguished from the leader of the community organization. The former was involved in expanding community organizing tasks in other villages, i.e. externally focused, while the latter led internal organizational tasks and activities. The LCOs in Bolinao were not employed by the NGO but rather considered as 'community scholars' and were given a small allowance for doing their tasks (Arciaga *et al.*, 2002). In Taytay, Palawan, Philippines, the LCOs were employed by the project implemented by an NGO and thus paid for their work. However, their tasks were similar to those implemented by the Bolinao LCOs.

Box 6.5. Local Community Organizers, Africa.

In Africa, different from Asia, many of the co-management programmes have been initiated and facilitated by government staff rather than NGOs. In South Africa, many of the co-management programmes (for example, Sokhulu Mussel, Klienmond Inshore Fishery, Kosi Bay Gillnetting, Amadiba Tourism) have been initiated and facilitated by provincial and national government officials. These individuals have acted as catalysts for partnership development with local leaders and fishers. This role has been particularly significant with respect to securing funds for project activities, lobbying for access rights, mediating disputes, implementing training initiatives and exploring alternative economic opportunities.

Source: Sowman *et al.* (2003).

Box 6.6. Case Study: Community Fisheries Facilitation in Pursat Province, Cambodia.

The term community facilitation is used in Cambodia to describe the community extension work of supporting organizations. Facilitation in Khmer is translated to *samrob samrourl*. *Samrob* means 'to make things go together in the same direction' while s*amrourl* means 'to make things easy'. Thus, s*amrob samrourl* is simply 'creating a process of doing things together in the same direction in an easy way'.

In Kampong Por commune in Pursat Province, the Cambodia Family Development Services (CFDS) has been working to improve the poor's conditions and spur their empowerment. Pursat is one of the six provinces surrounding Tonle Sap Lake. The programme is focused on helping villagers in the planning, organizing and facilitating of participatory local development, which is from the village to the provincial level. The programme is a strategic part of CFDS's vision of a Cambodian civil society of a fair, just and peaceful society, which will be achieved through an acceleration of economic growth to raise the living standards of all Cambodians.

The CFDS programme also supported a pilot project in developing a community fishery in Anlong Raing as a model of community fishery in Pursat. Until the change in government policy in 2001 towards fishery reform, there was little or no external support for the fishers and their communities. In addition, the outdated fishery law and lack of control and management of fishing lots, pollution and degradation of ecosystem and biodiversity has contributed to the heavy burden of most fishers for survival. Thus, the community did not benefit from the rich resources of Tonle Sap and currently they are being asked to take on the responsibility to manage their own community without building any capacity to take on the new task. The CFDS programme is an opportunity for them and the community to become better engaged with the Department of Fisheries in Pursat and the provincial officials looking for future collaboration in supporting the community fishery to build necessary knowledge, skill and attitude towards effective and sustainable management of resources in their community.

Sources: Cambodia Family Development Services, unpublished programme documents; Rivera-Guieb (2004).

The CO may be male or female depending upon the context of the community. One or more COs may live in the community. Often the external agent will provide technical support to the CO with other staff members who may have specialized skills. Many COs have a college degree in social work or community development, although those with other degrees make excellent COs. The external agent will need to train the CO on their philosophy of co-management and on methods and tools to be used. Many external agents have developed a specific co-management, community organizing or development process which they follow in their work.

The CO should have the following skills:

- Open-minded;
- Creative;
- Respectful;
- Sensitive to local culture and gender;
- Sense of humour;
- Modest and humble;
- Puts people at ease and does not set himself or herself apart or act superior;
- Facilitate and guide rather than lead the process;
- A clear understanding of the different theories of development;
- Familiarity with the concept and process of community organizing and participation processes;
- Social and community relationship skills such as skills in establishing rapport, conflict management and group maintenance;
- A clear grasp of community-based co-management concept and process;
- The ability to work with teams of professionals involved in the management of marine and coastal resource;
- A clear perspective of when to phase-out and to 'let go';
- Interviewing and documentation skills;
- Ability to facilitate group meetings and discussions;
- Communication skills (DENR *et al.*, 2001c).

Among the many skills and qualities a CO should have, the ability to dialogue – a fundamental aspect of community work – is one of the most crucial yet difficult tasks. Dialogue is an interchange and discussion of ideas based on a process of open and frank questioning and analysis in both directions between the community workers and the people. Community organizers cannot do their task if they do not take the time to listen to the people and constantly look for venues to interact with people. These may be in the form of meetings or house visits. The key element is to be able to dialogue with people in informal and intimate ways. A CO does not decide in advance what the community needs to know but dialogues with them to understand their needs.

The external agent should make logistical and administrative arrangements to support the CO. In many cases, the CO will need to travel to several communities or government offices. Resources will need to be made available for travel. Arrangements for communication through telephone and internet will need to be made for the CO. The external agent will need to provide back-up to the CO in case of sickness or other problems and in case a CO needs to be changed for any reason. In some situations, a CO may not be acceptable to the community and will need to be replaced. The external agent will need to make arrangements to pay the CO's salary and provide benefits.

Before entering the community, the CO should become familiar with the area, including its history, resources, culture, economy, social structure, problems, needs and opportunities. This information may be obtained from secondary data sources such as reports and publications and from key informant interviews with those knowledgeable about the area such as local elected officials, other NGO staff that may have worked in the area and government agency staff.

In conducting community entry, the CO should:

- Know the audience;
- Know the background of the community and its leaders;
- Initiate informal discussion with local government officials;
- Become acquainted with leaders and key informants;
- Know the potential topics they might want to discuss;
- Prepare appropriate techniques (for example, interviews and visualization techniques);
- Prepare secondary materials as background information.

6.2. Integration

In order to be effective, the CO will need to integrate himself or herself in community life in order to have knowledge about community members; their culture, livelihoods, institutions, social structure and environment; and their needs, behaviour and problems. This is a critical time for the CO as he or she must be accepted by the community. There must be credibility, trust and respect between the CO, the community members and government. This will involve long days and nights of listening, observing, talking and involvement in community activities. It will involve explaining the co-management programme and adapting to local conditions. The CO must become accustomed to local culture and traditions. This can take several months.

In order to gain this mutual trust and respect, the CO should observe, participate and hold informal conversations with a wide range of community members (Box 6.7). The CO should become a part of the community and participate in local economic and social activities. This may involve going to the fish landing site and talking to fishers, asking fishers to go fishing, talking to fish buyers and sellers in the market, attending local meetings, meeting women as they wash cloths, singing karaoke. It involves being available and observant.

> **Box 6.7.** Observation.
>
> Observations are qualitative descriptions of what is seen and heard and are obtained by attentively watching and recording the surroundings. Observations provide insight into activities that are difficult for people to describe and provide information on relevant activities, stakeholders and material culture. Directed observation looks at a specific activity, such as fish landing, or tries to answer a specific question, such as, 'How are cooperative meetings conducted?' Continuous observation seeks a broader understanding of activities and takes note of all regular activities throughout the day and night.
>
> Pay attention to everything and use all senses to observe. Introduce yourself and explain what you are doing. Ask questions concerning things relevant to the parameters being investigated, particularly activities which are not recognized. Take notes and take photographs (if possible). Fully record the activities taking place. Sketch as many things as possible. If possible, play an active role in the activity.
>
> Be aware that observation can sometimes be intrusive and may involve time from local people. Interacting closely with particular stakeholders can affect their activity and interactions with other stakeholders, particularly when there are conflicts between and within groups.

6.2.1. Courtesy call

The CO should make courtesy calls to elected and traditional leaders. Elected leaders include those at provincial/state, municipal and village government levels, for example, a provincial/state governor, a municipal mayor or a village council person. Traditional leaders include village chiefs, village elders, religious leaders and senior fishers. The purpose of the courtesy call is for the CO to:

- Introduce himself or herself;
- Introduce the programme and its objectives;
- Introduce the concept of co-management;
- Introduce the approach to be taken to community participation and activities;
- Determine the role of elected and traditional leaders and government;
- Open communication and dialogue;
- Determine the level of support;
- Encourage participation and cooperation; and
- Determine the needs of the leaders (government and traditional).

This should not be a one time visit but the start of a long-term relationship. Depending upon their interest, the leaders should be actively consulted and brought into the co-management programme. There may be a need for frequent visits to inform and discuss programme activities and needs.

It is important to note if the leaders are male-dominated; it is the role of the CO to consciously seek out and consult women members of the community, even if they do not play formal leadership roles.

At this early stage, the CO may also make courtesy calls on government offices and staff for the same reasons as the leaders. This may include, for example, provincial/state and municipal fisheries office, environment and natural resource office, economic development office, agriculture office and local government office.

The significant role of government and traditional leaders in co-management should be recognized. If any leaders show lack of or limited support, consider having someone from the community or someone who is known or respected by the leader and is supportive of the project visit with the leader and introduce co-management and the programme. Consider consulting with other leaders or government officials who could provide alternative channels to gain cooperation of the non-supportive leaders.

6.2.2. Community meetings

In a similar fashion as the courtesy calls on government and traditional leaders, the CO needs to inform the community about the programme. Community meetings are one method of doing this (Box 6.8). The community meetings should have the same general purposes as the courtesy calls, that is:

- Introduce himself or herself;
- Introduce the programme and its objectives;
- Introduce the concept of co-management;
- Introduce the approach to be taken to community participation and activities;
- Determine the role of community members;
- Open communication and dialogue;
- Determine the level of support;
- Encourage participation and cooperation; and
- Determine the needs of the community members in terms of training and awareness raising.

These community meetings can be part of previously organized meetings, such as government, social or religious gatherings. They can be meetings organized by the CO specifically for the purpose of informing community members about the programme. Or these meetings can be informal where the CO meets with groups of people in daily settings such as the market or fish landing site. The CO should hold as many meetings as needed to ensure that community members are informed and aware of the programme, are familiar with the CO and know how to contact the CO.

The meeting agenda should include an introduction of people associated with the programme (CO, government and community leaders, external agent), a presentation about the programme and open discussion about the programme and any inputs and concerns from the meeting participants.

The CO should have a formal presentation prepared which covers the above topics about the programme. A formal presentation will ensure that the same information about the programme is presented at each meeting.

Box 6.8. Community Meetings.

Community meetings can be one of the most important tools for community information gathering and for the communication of information. Community meetings can serve a wide variety of purposes:

- Give and receive information;
- Discuss issues of relevance;
- Gain a consensus on an issue;
- Identify problems and solutions;
- Plan activities, negotiate conflicts;
- Validate interpretations of studies and evaluations.

A community meeting generally involves a large number of people but, if well designed, can be participatory by encouraging two-way communication. Smaller focus-group meetings can be even more participatory as the information sharing may be more equitable when there is a particular issue or purpose or when the group members are comfortable speaking to each other. The output of focus-group meetings can be presented to larger group meetings, giving a 'voice' to those in the community who do not feel comfortable speaking up at large group settings. Regular small group meetings can foster a cooperative approach to decision-making and discussion.

A lot of careful planning goes into a successful meeting.

- Have a clear purpose and know what is to be accomplished.
- Obtain the approval and involvement of local leaders and be aware of community protocol.
- Arrange a time and place for the meeting, taking into account the activity patterns of the participants. The time and place of a meeting can encourage or discourage attendance.
- Check whether food and accommodation will be needed, especially for outsiders.
- Inform the community of the meeting through posters, public announcements, radio, word of mouth.
- Plan and prepare handouts and material to be distributed and presented.
- Use the language of the community.
- Plan smaller, more limited group meetings (if necessary) and develop feedback mechanisms.
- Skilled facilitation is crucial. Identify a skilled and respected person to facilitate the meeting.
- Plan the meeting process to encourage discussion. Think of two-way communication and how to encourage participation. Prepare some leading questions to encourage discussion.
- When facilitating the meeting:
  - Make the purpose of the meeting clear in the introduction and explain the process of the meeting.
  - Prepare and check all presentation equipment and materials.
  - Begin and end more or less at the stated time.
  - Start with issues on which it is easy to reach an agreement or to accept differences of opinion.
  - Allow conflicting opinions to emerge and try to have these differences either resolved, or accepted by the group.
  - Have a recorder keep notes of the meeting.

> ○ At the end of the meeting, summarize the proceeding, outline the decisions that have been made, and identify next steps.
> ○ If appropriate, confirm the time and place of the next meeting.
>
> Consider that there may be factions of the community that are unable or unwilling to speak out at a large group meeting. Separate meetings with these people can be held, and their perspectives can be brought back to larger group meetings. Large group meetings are not an appropriate venue for resolving conflict and this should be done in a smaller group setting. Several meetings may be needed to obtain the desired results.
> Beware of hidden agendas and groups who may want to use the meeting for their own purpose. The facilitator must have enough authority to keep the meeting on track and enough sensitivity to include people in the discussions. The group may see the facilitator as the 'expert' to carry the whole meeting. The facilitator should keep handing the questions back to the group.
>
> Source: Townsley (1996).

Published materials about the programme (including purpose, objective, partners, role of community, co-management) should be available. The presentation and published materials should be easily understood and in the local language.

As an alternative method, and as will be discussed below, the CO may need to conduct a household census. In doing so, the CO will have the opportunity to meet with community members at each house and provide an explanation of the programme.

6.2.3. Key informants

The CO should identify key informants in the community that can provide reliable information on a variety of topics and issues (Box 6.9). There may be one or more key informants for each topic and issue. Key informants are purposely selected community members who are able to provide information on a particular topic based on their knowledge, skills or experience with that subject. It is felt that community members can provide accurate, relevant and detailed information about their community.

6.2.4. Preliminary community profile

At this early stage, the CO may want to develop a preliminary community profile including information on community demographic characteristics (population, age, sex, education, occupation), history, conflicts, and problems and needs. This will allow the CO and others, such as the core group, to have a better understanding of the community and how to direct activities. Some of this information may have been collected in the 'beginnings' or pre-

Box 6.9. Key Informants.

Different types of people have different types of knowledge. If you ask the same question to a child, a woman, a less educated person and an older man, you may get four different answers. The type of knowledge people have is related to their age, sex, education, occupation, social status and history.

In working with key informants, first identify the type of information that is needed. Ask a village leader to identify individuals in the community that hold key positions or are widely respected. Alternatively, ask a broad sample of people to name others in the community who know most about a topic. Choose who among these people can provide relevant information based on the information needed. Meet key informants in a place convenient to them and that allows for undisturbed discussion. Begin with open-ended questions (questions which encourage follow-up discussion) and become more specific through the interview. Ask the key informant to identify someone else that could be helpful in giving information about the topic.

Source: IIRR (1998).

implementation phase. A more detailed community profile will be conducted under the research activity (see Chapter 7, Section 7.4).

Community demographic characteristics are available from census information. Current census information should be available at provincial/state and municipal government offices. In most cases, the village council secretary also holds the village census and other information. A census is usually conducted every 10 years. Some municipalities collect certain demographic information (such as population, birth and death rates) on an annual basis. In the beginnings phase of the programme, a preliminary problems and needs identification may have been conducted. If so, this will provide the CO with valuable information to direct activities.

If a research or development project has been conducted in the community, information on community demographics and problems and needs may be available in reports or publications of the project; for example, an academic thesis from a local university or work done by an NGO. The CO should ask government staff and local people if they are familiar with any recent studies and seek them out to supplement census data.

If information on community demographics and problems and needs is not available, is out-of-date or is questionable, the CO may want to conduct a household census. This is more easily done if the community is small. A short, one-page, household census can be conducted relatively quickly and inexpensively. The census is conducted for every household in the community. (Alternatively, a census could be conducted of only fishing households by asking a first question on whether fishing is an occupation of the household and surveying only those households that answer yes to this question.) It would allow the CO to have a current community profile, target specific information on households (for example, location of all fishing households), allow the CO to visit and introduce himself or herself at all households, and gain an understanding of problems and needs. The census form should be short and may include questions on:

- Household size;
- Age of each household member;
- Sex of each household member;
- Education of each household member;
- Religion of household;
- Ethnicity and origins of household (migrants usually have different social networks and dynamics in the community);
- Primary and secondary occupation of household;
- Problems and needs;
- Member of a community organization;
- Which organization;
- If fishing household, type of fishing gear used.

The census questionnaire should be prepared to be conducted in 10–15 minutes. If the community is small enough, the CO should be able to conduct the census alone. This will allow the CO to walk the whole community, observe and meet people. The census data can be analysed and reported in tables.

The CO should prepare a short history of the community. The history can include when the community was formed, population and other demographic trends over time, political history, major conflicts and resource use patterns. This information can be obtained from secondary sources, community leaders and the core group members. This report should be short as more detailed information on the community will be prepared as part of the research activities of the implementation phase.

A preliminary community profile report is prepared and shared with leaders and the community as an initial output. The report describes the community in general terms and can serve to direct initial co-management planning and activities.

6.2.5. Area boundaries and management unit

A preliminary identification of resource management boundaries should occur. These boundaries will serve later to establish the resource management unit during the co-management plan activity. Thus, the emphasis here is on mapping boundaries to serve as information for later planning. The boundaries should make sense from a biological or ecological standpoint (they would cover the fishing area of local fishers or the essential elements of the coastal ecosystem), from a social standpoint (they would cover political jurisdictional boundaries, as well as traditional tenure areas) and from an economic standpoint (they would ensure that benefits are enjoyed by those who bear the costs).

An effective management unit would comprise a series of nested management units composed of different scales. The complexity of natural and social systems requires that management be addressed simultaneously at different scales. Ecosystems and social systems tend to be organized hierarchically. Each level in the hierarchy is independent, to some degree, on the levels above and below, but also dependent on these levels. Co-management institutions involve horizontal (across space, i.e. networks of communities involved in fisheries management) and vertical linkages (across levels of institutions or organizations, i.e. multistakeholder bodies or networks of government agencies).

The management unit should not be so large as to be unmanageable, nor so small as to be ineffective. It should be large enough to accommodate an ecosystem or habitat, and small enough to accommodate a management unit in charge. The management unit may be started from a small and clearly defined geographic area or ecosystem, or from a recognized social unit (i.e. local government) and its management area.

Local community members have a great deal of knowledge about characteristics of boundaries and features and should play active roles in this preliminary and later boundary identification. Resource mapping is a useful participatory technique to identify boundaries (Box 6.10).

There can be several types of boundaries: political, ecological, fishing tenure, fishing gear area, planning and management. These should be identified. The boundaries may be different but overlap.

Political boundaries represent the local government's authority over land and waters. Coastal terminal points on land must be determined and validated before delineating political boundaries. Agreement between adjacent local governments is necessary to finalize the coastal terminal points. National government authority defines the local government's authority over waters.

Ecological boundaries represent the aquatic ecosystems in the area. If no resource and ecological assessment information is available, it may be possible to identify the ecological boundaries in a preliminary way through resource mapping with local fishers and other resource users. Traditional fishing tenure areas and fishing gear use areas can also be identified through resource mapping with fishers.

Planning and management boundaries are usually based on specific issues or problems that are addressed later in the planning process. A planning boundary should be sufficiently inclusive to ensure that important impact-generating uses and activities are included, but not so large as to dilute the programme. A management boundary denotes the area within which specific regulatory, developmental or other management activities are designed to occur in order to reduce adverse impacts on coastal resources, reduce risks of hazards or increase opportunities for optimal resource use. Planning and management boundaries should be developed and validated with community participation and input during the plan preparation (DENR *et al.*, 2001b).

Boundaries need to be identified on a map. A base map is prepared which shows general political and geographic/physical features to serve as reference points. The base map can contain new information and serve to map management strategies. If available, a topographic map and a nautical chart can provide useful information. A global positioning system (GPS) electronic device can assist in identifying boundaries and locating important features.

Box 6.10. Resource Mapping.

Resource mapping is a participatory method that allows community members to identify, locate and classify past and present resource occurrence, distribution, use, tenure and access, and to reveal the significance the participants attach to them. It can allow the establishment of relations between information sets and their spatial location. Resource mapping can apply to all ecosystems known to the community and the scale of the maps can be set/adjusted depending on the required level of detail.

A group of participants or key informants are invited to a mapping exercise. The group may be stratified or grouped by age, fishing gear type or gender depending upon the needs of the resource mapping. Prepare a base map of the area showing general boundaries of the area so that participants can have reference points (alternatively, the participants can prepare the base map, identifying boundaries and features). Collate a checklist of resources or features to be mapped. Consider that only a limited number of topics can be mapped. Position the paper so that it can be seen by all. Ask the participants to locate on the map the listed resources and features. Allow for additions as needed. Use symbols and colours to represent various sets of information and generate a corresponding legend. Validate the findings. A facilitator should guide the discussion and a single documenter should write on the map. Make copies of the map when completed and share with a wider forum. Repeat as necessary for other resources and features.

Variations of resource mapping include:

1. Stratified resource mapping which involves dividing participants into groups according to gender, age, ethnic origin or other categories. This is useful in identifying relationships of social groups and resources.
2. Gendered mapping highlights men's and women's access to, control over and perceptions regarding the importance of certain resources. Gendered mapping is usually conducted among separate groups of men and women.
3. Two-stage resource mapping involves transposing the information from the sketch map to a conventional topographic map or nautical chart. Two-stage resource mapping may be used by the community in dealing with formal institutions on particular issues related to tenure, usage rights, right of way, etc.

Source: IIRR (1998).

Where possible, base maps and other spatial data should be put in geographic information systems (GIS) to serve as an important starting point for planning, in the same manner as land and sea use maps are developed (DENR *et al.*, 2001b).

6.2.6. Problems, needs and opportunities identification

If a household census is not conducted or if no current information on community problems, needs and opportunities (PNO) is available, the CO will need to begin to identify community PNO. This PNO identification drives the co-management programme as it provides focus for developing the programme action plan, strategy and activities. PNO identification allows the community, working with the CO, to define, analyse and rank community problems, needs and opportunities. This is done according to the importance, the urgency of finding solutions, the number of people affected, and the probability of resolution or taking advantage of the opportunity through community action. A PNO identification may also identify new or different problems and needs which were not anticipated or earlier discussed.

The CO can assist in identifying problems, needs and opportunities through community meetings and group discussions. The idea is to identify, group and rank problems, needs and opportunities in order of priority. There are several methods which can be used for PNO identification (Boxes 6.11, 6.12, 6.13 and 6.14). These methods should encourage people to confront their own problems and needs.

Box 6.11. Prioritizing Problems, Needs and Opportunities.

One possible method for problem, need and opportunity analysis and prioritizing is *criteria ranking* (IIRR, 1998):

- While it may be possible to work with the entire community at one time, it is often impractical.
- Identify manageable size groups within the community to work with. While they do not have to be homogeneous, they should have common interests.
- Identify a time and place for a meeting. It should be a safe and relaxed atmosphere that will allow for open and free discussion.
- Explain the objective of the meeting and begin with a general discussion on the project and community.
- Get the members of the group to brainstorm and suggest both community and fisheries (or coastal resource) problems and needs. Ask them to write these on pieces of paper and put them on a board.
- Get the group to classify the problems and needs into related groups.
- Suggest and explain possible criteria for ranking the problems:
  - Extent or scope;
  - Degree of impact on a particular resource;
  - Occurrence or regularity;
  - Other criteria may be suggested.
- Let the participants analyse each problem according to the criteria set.
- Ask the participants to compare each problem. Score each problem using a predetermined scale (e.g. 1–5, where 5 is the highest score).
- Add the total score for each problem and place the sum in the second to the last column. The sum reveals the relative importance of the problem across all criteria and determines how it ranks compared to the other problems.
- Use the last column to rank the problems.
- Some of the descriptions may be qualitative, e.g. degree of impact and occurrence. Ask for an explanation and clarify meanings.

Another useful and rapid prioritizing method is called *pairwise ranking*. Pairwise ranking is used to determine the main preferences of individuals, identify their ranking criteria and easily compare the priorities of different individuals.

Pairwise ranking generally involves the following steps (adapted from Margoluis and Salafsky, 1998):

- Begin by identifying all of the goals (and/or associated objectives) to be prioritized.
- Then record each of these items on a separate index card with a pen.
- With the help of those involved in the ranking exercise, list out each possible pairing between any two of the goals/objectives in your set.
- Next, working down the list of possible pairings, place the first pairing (index cards) of goals or objectives in front of each respondent and ask her/him to choose the more important goal/objective between the two. Record the choice in a table. Ask the respondent to explain why s/he made the choice and record the response in a second table. Alternatively, a focus group can act as a single respondent if it is able to come to consensus on choices.
- After this, present and choose between the next pair of choices, recording the response and reasoning. Continue through the list until all possible combinations of pairings have been chosen with the respondent and the table of responses has been completed for the individual.
- Without the respondent looking on, quickly tabulate and score the overall preferences by counting (sum) the number of times each item was chosen over any other. Record these scores in the table and list out the most preferred (highest score) to least preferred (lowest score) goals/objectives based on the total score.
- Crosscheck the results by asking the respondent what s/he thinks is the most important preference out of all of the potential goal/objective choices.
- Repeat the exercise for the necessary number of respondents.
- Finally, tabulate total preference scores across all respondents to determine the group's overall preferences. List these overall preferences from highest to lowest and begin to discuss how to divide the ranked items into higher and lower priorities.

When identifying community problems, needs and opportunities, listen to the community. The community knows its problems, needs and opportunities. Formulate the PNO in a clear and specific manner. As a CO and facilitator, consider what your capabilities are and think about which problems should be solved first. Be honest and realistic about what can be done. The CO should be guided not by what is important to them but by what is important to the community and what can be handled now (Rivera-Guieb and Marschke, 2002).

The output of the PNO identification should be a report made available to all stakeholders. The report should be reviewed and verified with the stakeholders. The report may be revisited before the co-management plan and strategy is prepared since time may pass between the activities. The PNO identification activity should be conducted on a regular basis to evaluate changing problems and needs in the community.

6.2.7. Identify potential leaders

The CO should identify candidates in the community who could become leaders in the co-management programme and help in their training and

Box 6.12. Problem Trees and Solution Trees.

Problem trees are diagrammatic presentations of a problem, its cause and effects. These are done after a community has identified and prioritized its problems. The focus is on one problem, its cause and effect, at a time. The problem tree can serve as the focus for developing a plan of action to resolve the problem or need.

- Draw a large tree on a board with the problem on the trunk.
- Let the participants brainstorm over the causes of the problem by asking the question 'why?' Draw a root for each cause, and write the cause on the root.
- Repeat the question 'why?' for each cause to identify secondary causes. Write these lower down the roots below the primary causes.
- Ask the participants to identify effects or impacts of the problem by asking 'what happened?' Draw a branch for each effect and write the effect on the branch.
- For each effect, repeat the question 'what happened?' to reveal secondary effects. Place them higher up the branch.
- Continue until the participants can no longer identify any effects of the problem.
- Repeat for other problems.
- Participants can write all the causes and effects they identify on separate cards and pin them on the tree. Participants can then easily move cards around as they see links among the causes and effects.
- In a similar fashion, a solution tree can be developed by the group to identify strategies toward overcoming the problem and achieving a better outcome.

Source: IIRR (1998).

Box 6.13. Brainstorming.

Brainstorming is a group activity where selected members of the community take turns in sharing their ideas relating to a certain topic or question. Brainstorming encourages participants to think critically and creatively rather than to simply generate a list of options, answers or interests.

- Set the objective of the activity (e.g. problems and needs identification).
- Based on the objective, determine the individuals and groups that should be involved in the activity.
- Invite the participants and set a time and place.
- Explain and discuss the objectives and mechanics of the session so that everyone understands the process.
- Introduce the topic.
- Ask each participant to give or share their ideas relating to the topic.
- Write down all ideas on a board or paper.
- With the participants, sort, classify and synthesize ideas written on the board.
- Discuss and analyse the results.
- In problem identification, ask the participants: Who is affected? How many people are affected? How? How does the problem affect the community? How do community members feel individually and collectively? Which problem is the most urgent? Which problem is the most manageable to try to solve, considering group resources and limitations?

- The problems or ideas can be classified, segregated or synthesized. They can also serve as the starting point of a new brainstorming session.
- An alternative when participants are hesitant to speak out is to use idea cards. Cards are distributed to each participant and they are asked to write their ideas on the card. The cards are collected and put up on a board and discussed. The cards can be moved around and grouped on the board as needed.

Source: IIRR (1998).

Box 6.14. Guidelines for Facilitating Focus Group Discussions.

Focus group discussions are small discussions with four to eight selected members of the community who are chosen for their knowledge or involvement in a specific issue.
    The approach involves:

- Establishing the objectives of the discussion;
- Selecting participants based on knowledge and involvement in the issue;
- Planning the timeframe for the discussion;
- Designing focus group guidelines and following guidelines for leading group discussions;
- Questions should be phrased to discover the community attitudes and perceptions about the issue.

    The following are some standard operating procedures for facilitating focus group discussions:

- Always begin by introducing the facilitator and participants;
- Start each session with a cultural ritual or prayer if appropriate for the group;
- Make sure the language used is understood by participants or use a translator;
- Start the session by explaining the objectives, describing the agenda or activities and identifying the desired outcome;
- Explain the process the group will go through, the roles of the participants and the expected timeframe;
- Have someone besides the facilitator document the discussion and outputs in meeting minutes and give a copy to the group;
- Always include the names of participants and date on any output;
- Be resourceful and creative and use interesting audio-visual aids;
- Be sensitive to participants' needs; take breaks when needed; allow the agenda to change if other important issues are raised;
- Choose an appropriate time and place for the community to participate;
- Do not rush the participants; work at their speed;
- Encourage participation by all, control participants who dominate the group;
- Listen carefully to participants and do not interrupt;
- Settle disagreements through dialogue and consensus-building, exhaust all arguments until a resolution is reached;
- Be gender- and culture-sensitive and create an environment of respect.

Source: IIRR (1998).

preparation. Leaders may already have stepped forward in the 'beginnings' stage, and others may come forward later. The CO may want to identify other candidates to develop diversity in leadership and to serve specific activities. Fishers tend to be independent people but co-management requires leaders to lead the process and the organizations which are involved in co-management. Leaders are needed to motivate, inspire, manage and energize the organization and process (Box 6.15). Leaders may have some natural ability to lead, but leadership can also be learned through experience (Box 6.16).

Box 6.15. Long-term 'Champions'.

In cases of co-management in South Africa, an important role was found to be played by one or two dedicated persons intimately involved with a project. A project 'champion', whether in the community, in the responsible management agency, or in an external NGO or academic institution, was key to motivating partners, encouraging commitment and providing continuity and support to the partners during the ups and downs experienced in the planning and implementation of the co-management arrangements. These 'champions' or mentors frequently facilitate communication and interaction between communities and relevant government departments and other stakeholders and broker the co-management arrangements. They also play an important role in keeping local users up to date on relevant legislative, administrative and political changes and initiatives.

Source: Sowman *et al.* (2003, pp. 317–318).

Box 6.16. How to Spot Potential Local Leaders.

- Conduct a community study and develop a community profile;
- Immerse yourself among the people;
- Observe people's activities and be willing to learn from them;
- Observe individual people's activities and their relationship to people;
- Build rapport with initial contacts;
- Ask trusted persons to vouch for contact persons;
- Take advantage of 'message bearers' and 'go-betweens' when communicating with people.

Source: Almerigi (2000).

A good leader:

- Is willing to share power with the group;
- Listens;
- Communicates;
- Collaborates;

- Expresses the values of the organization;
- Pulls together strengths and talents of the group;
- Has a caring and positive attitude towards people;
- Values consensus;
- Is open-minded, flexible and willing to compromise;
- Leads by example; and
- Considers the group over self.

A leader is a steward, or caretaker of an organization, one who has been entrusted to work for the benefit of all (Almerigi, 2000).

6.2.8. Assessing existing organizations

As will be discussed further, a community organization(s) is one of the foundations of co-management. Existing organizations in the community are identified and assessed to determine their potential to participate in the co-management programme.

Organizations are groups of individuals bound by a common purpose to achieve objectives (North, 1990). Organizations can be formal (i.e. those registered with government) and informal (i.e. those that are traditional). Organizations may be called cooperatives, associations, councils or non-governmental organizations, and may exist for a number of purposes including resource management, fishing, religious, youth, women, sport, marketing, etc. An organization may have a formal management body and rules or may be an informal grouping of individuals for a specific purpose. In co-management, community organizations and their representatives participate in decision-making and managing and protecting the fisheries and aquatic resources. Before deciding to form a new organization, the CO should identify and assess the existing organizations in the community. Organizations may exist in the community that are already capable of engaging in co-management or just need to be strengthened. In some cases there may be existing organizations providing services or operating projects in the community. If these services can be useful to the project, the CO should consider the possibility of executing a memorandum of agreement with the organizations so as to avoid duplication of services and interventions.

Secondary data and the community census can provide information on organizations in the community. The CO may also want to interview community leaders and key informants to identify community organizations. Formal organizations may be easier to identify than informal organizations. It will be important to talk to a wide range of individuals in the community to identify all organizations.

For each organization the following information should be obtained:

- Name;
- Address/location;
- Purpose;
- Years in existence;

- Formal/informal;
- Contact person;
- Names of management members (president, vice-president, treasurer, secretary, board members);
- Number of members (male/female);
- Annual reports;
- Strategic plan/rules/regulations;
- Meeting schedule.

Several factors to be considered in assessing existing organizations include:

- Does the organization represent the different sectors in the community (e.g. youth, fishers, women, etc.)?
- Was the organization formed to address issues related to co-management or resource management?
- Does the organization have a mechanism to deal with dynamics/issues within the group (e.g. interpersonal conflicts, delineation of roles and functions, etc.)?
- Does the organization have a legal personality and/or credibility in the bigger community?
- Is the leadership structure and pattern democratic enough to promote maximum participation from members?

For each organization, a profile with the above information is prepared which will assist in assessing the organization's capability to engage in co-management.

6.2.9. Formation of core groups

A core group is a small group of individuals from the community (perhaps four or five) who will initially work with the CO to guide the co-management programme. The core group should only operate until viable and functional community organization(s) and a co-management body are in place (see Chapter 10). The members of the core group should represent different sectors of the community. The core group is crucial as it gives initial real responsibility and power to the community members for management. The core group will be:

- A highly cohesive and committed group;
- The basic building block of the organization(s);
- A training ground for leadership;
- A forum for practising people's participation; and
- Initial co-management management team (Almerigi, 2000).

Depending upon the size of the community and the context, there may be several core groups established to address specific co-management activities or issues.

The core group can serve to:

- Facilitate the circulation of information among community members;
- Develop dialogue and provoke social discussion about community and resource management issues;
- Facilitate community organizing;
- Identify problems, issues and opportunities;
- Assist in programme decision-making;
- Identify stakeholders and stakeholder groups;
- Assist in the gathering of information.

The core group members may come forward by themselves and volunteer to be involved based on personal motivation, such as the initial leaders identified earlier, or they may be selected by the CO and/or community leaders. Depending upon the situation, the CO may be a member of the core group.

The following criteria can assist in choosing the core group members:

- Be credible to community members (either certain community groups or the majority of the community);
- Be accessible to community members;
- Represent a diversity of community interests and groups;
- Be well-respected;
- Be conscientious and resourceful in their work;
- Have good communication skills;
- Be open-minded and desirous of change (Almerigi, 2000).

Other criteria may be added for a specific context and community.
Some key qualities of a good core group are:

- Being active, efficient, fair, multi-disciplinary and transparent in decision-making;
- Acting on the basis of consensus and collaboration;
- Being determined to launch but not to lead or dominate the co-management process (Borrini-Feyerabend *et al.*, 2000).

It can be expected that there will be differing levels of knowledge and experience among the core group members about co-management and the various activities associated with it. The CO will need to provide information and training to the core group on specific topics and issues, such as the co-management programme and various activities, as needed to assist them in their work. The core group may need to meet regularly during the initial phases of implementation to make decisions and guide the programme.

If core group members are hesitant to join in the co-management programme, consider having informal discussions with community members to find out probable causes why they are unwilling to participate. Consider any past experiences with co-management or development projects in the community and success or failure.

6.2.10. Identifying stakeholders

There are many potential stakeholders in community-based co-management. Stakeholders are individuals, groups or organizations of men and women, old and young, who are in one way or another interested, involved or affected (positively or negatively) by a particular project (Box 6.17). They may be motivated to take action based on their interest or values. Stakeholders may include groups affected by the management decisions, groups concerned about the management decisions, groups dependent upon the resources to be managed, groups with claims over the area or resources, groups with activities that impact on the area or resources and groups with, for example, special seasonal or geographic interests. Stakeholders are important because they can support or not support co-management, which can lead to its success or failure. Stakeholders of coastal resources include fishers, fisher households and fishing communities, but also include seasonal fishers, boat owners, fish traders, government, mangrove cutters, etc. Stakeholder groups can be divided into smaller and smaller sub-groups depending upon the particular purpose of stakeholder analysis. The identification of key stakeholders should be inclusive and detailed. More groups may mean more problems and discussion, but excluding certain groups could lead to problems in the long run. Ultimately, every individual is a stakeholder, but that level of detail is rarely required (IIRR, 1998). A key question to be answered in the co-management programme is:

Box 6.17. Identifying Potential Stakeholders.

- Are there communities, groups or individuals actually or potentially affected by the management decisions?
- Who are the main traditional authorities in the area at stake? Are there government agencies officially responsible for the management of the area? Are there respected institutions, to which people have recourses in connection with a variety of needs and circumstances?
- Who has access to the marine resources at stake? Who is using the resources? In what ways? Has this changed over time?
- Which communities, groups and individuals are most dependent on the resources at stake? Is this a matter of livelihood or economic advantage?
- Who upholds claims, including customary rights and legal jurisdiction over the area? Are there communities with ancestral and/or other types of acquired rights? Are various government sectors and ministerial departments involved? Are there national and/or international bodies involved because of specific laws or treaties?
- Which communities, groups or individuals are most knowledgeable about, and capable of dealing with, the resources at stake? Who has direct experience in managing them?
- What are the seasonal/geographic variations in resource use patterns and user interest?
- Are there co-management initiatives in the region? If so, to what extent are they succeeding? Who are the main partners?

Source: Borrini-Feyerabend *et al.* (2000, pp. 22–23).

who are the stakeholders that are entitled to take part in discussions and in management?

Stakeholder analysis is conducted to identify potential partners for co-management (Box 6.18), to explore possible approaches in relating to a particular

---

Box 6.18. Stakeholder Analysis.

The process of identifying stakeholders and figuring out their respective importance regarding the resource is referred to as a stakeholder analysis.

The stakeholder analysis is best conducted in a participatory way with the core group and/or key informants (knowledgeable or important individuals in the community) from a single stakeholder group or from various stakeholder groups. The participants in the stakeholder analysis exercise need to be documented to objectively analyse the results.

- The resource, activity or project to be analysed is identified based on the PNO assessment (for example, fishing) (see Section 6.2.6). The objective(s) and intended outputs of the stakeholder analysis are identified.
- An understanding of the context of the resource, activity or project and how the overall system operates is undertaken including who are the main decision-makers in the system and interactions and interdependencies, especially ecological and human systems.
- The participants are asked to identify and list all stakeholders associated with the resource, activity or project. Write their names on paper circles. Larger circles are used to identify stakeholders with greater influence or power. (Alternatively, a focal group approach can be used by identifying a stakeholder group which plays a central role in resource use and management. Other stakeholders are then uncovered by identifying individuals, groups and institutions who have important relationships with the focal group with respect to resource management.)
- After an initial set of stakeholders is identified, they need to be verified. The stakeholders are questioned as to whom they perceive the other main stakeholders to be, and what the relations between different stakeholders are.
- Prepare a stakeholder analysis matrix with two columns (positively affected and negatively affected) and two rows (directly affected and indirectly affected). Place the circles in one of the four boxes on the matrix. Draw lines between the stakeholders to indicate the existence of some form of relationship. Use plus and minus symbols to indicate the nature of the relationship. An explicit investigation of the relationship between stakeholders can reveal information about the nature of conflicts and cooperative action, and the reasons and contexts behind them.
- Prepare a stakeholder analysis and coping matrix. This matrix has five columns:

1. Stakeholder group.
2. Describe the potential impact of the proposed action (for example, actions identified in the co-management plan) on the stakeholder group.
3. Describe the potential reaction of the affected group and the implications for the proposed action.
4. Can the proposed action be modified to reduce or mitigate the negative impact? If so, how?
5. Describe the recommended course of action (coping strategy).

Begin with the stakeholders identified as being directly and negatively affected, then move on to those indirectly and negatively affected, and so forth. Write the information to each question on the column of the matrix.

- After filling in the matrix, have the group discuss issues, problems and opportunities. Formulate courses of action for addressing various stakeholder interests, especially for those negatively affected.
- Repeat for other resource, activity or project.

This exercise can identify different stakeholders, give ideas on how to relate to particular stakeholders, and provide insights into the dynamics and relationships of different stakeholders. As a follow-up, the interests, characteristics and circumstances of each stakeholder group in relation to the resource can be investigated.

Stakeholder analysis should be done before initiating co-management. It should be repeated at key points in the co-management process to check on possible changes in the number and characteristics of the stakeholders.

Sources: Grimble and Chan (1995), IIRR (1998).

person or group who can be supportive or potentially hostile to co-management and to provide insights into the dynamics and relationships of individuals and groups with various interests in a particular resource or project. The stakeholder analysis is usually done by key informants from primary stakeholders. In some cases stakeholders may be easily identified through existing organized groups, while in others they may not be organized to currently engage in co-management (for example, a group of unorganized fishers using the same fishing gear type or fishing in the same area). The identification of not only groups with special interests and concerns but also those possessing specific capacities and knowledge for management can improve the list of key stakeholders.

Once key stakeholder groups are identified, it is important to find out their interests and concerns, whether they are organized and capable of participating in management decision-making, and whether they are willing to participate. Those groups that are not organized or prepared to participate will require assistance to do so.

If a variety of stakeholders is identified in the stakeholder analysis, which will probably happen, the question arises as to who should be invited to participate in the co-management programme (Box 6.19). This can create a dilemma. While it is important to have a well-represented co-management programme, it is important to determine if all stakeholder sub-groups are entitled to be involved in the programme. Too many stakeholders can create administrative and resource allocation problems. It is important that the final stakeholders involved in co-management be well-balanced; not too many so as to complicate and slow down the programme and not too few so as to leave out some key stakeholders. As such, the issue of entitlement becomes a central question: 'Who is entitled to participate in co-management?' This question needs to be addressed initially by the core group, and later by the co-management body (Box 6.20). It is difficult and is often only accomplished

Box 6.19. Identification of Bona Fide Fishers, South Africa.

A significant problem in South Africa is the illegal harvesting of abalone and rock lobster along the southwest coast. A project was initiated in 1999 to bring all the conflicting stakeholders together to identify and implement a coordinated strategy to diminish poaching in the Hangklip-Kleinmond area. It was necessary to identify the 'real' fishers for the project. The decision was taken at the outset of the project that the identification of the fishers in the community would take place by the community itself. However, the criteria for making this decision, and for evaluating the decision, were not effectively defined by the project. As a result, conflict emerged as to the credibility and legitimacy of people applying for, and receiving, access to resources.

First, one of the groups of fishers usually did not participate in meetings and workshops. Their commitment to the process and their ability to fish the quota, if allocated to them, was never verified. Second, there were questions raised regarding people potentially benefiting from the process if they had been outside of the fishing industry for many years doing other employment. Finally, some of the leaders themselves had not fished for many years and had also been involved in other professions. This caused concern as to whether they were bona fide fishers, or whether they were businessmen with other priorities and interests. There was never finalization of the criteria to determine who qualified as a 'fisher' with the community. Therefore, when rights were allocated after the project was terminated, conflict broke out within the community as fishers argued that some of the quotas did not go to the 'real' fishers in the community. The identification of and agreement on criteria regarding who qualifies as a 'fisher' needs to be clarified for future resource allocation processes.

Source: Hauck and Hector (2003).

Box 6.20. Examples of Characteristics of Entitlements.

- Existing legal rights to resource, whether customary or modern law;
- Mandate of the state;
- Direct dependency on resource for subsistence and survival;
- Dependency for economic livelihood;
- Historical, cultural and spiritual relationship to the area;
- Continuity of relationship, e.g. residents versus visitors and tourists;
- Unique knowledge about and ability to manage the area;
- Proximity to the area;
- Degree of effort and interest in management;
- Loss and damage as a result of the co-management;
- Number of individuals or groups sharing the same interest or concern.

Source: Borrini-Feyerabend *et al.* (2000).

through participation from and negotiation with groups and individuals to ensure equitable representation in the co-management programme. All who believe themselves stakeholders should be allowed to argue their case for entitlement. The stakeholders with recognized entitlements may be subdivided between 'primary' and 'secondary', and accorded with different roles, rights and responsibilities in co-management (Borrini-Feyerabend *et al.*, 2000). For example, full-time fishers may be recognized as primary stakeholders and seasonal fishers may be recognized as secondary stakeholders.

A problem to note on entitlements is that these are often based on legal and state-mandated rules and documents. In some cases, there are community members, the indigenous peoples among them, who cannot argue their rights to the resources within the legal framework. They would have their own traditional system of rights and rules. It is important for the CO and the core group to recognize the varying bases for claiming rights or entitlements to resources. Establishing equitable representation among stakeholders is necessary in a co-management process.

6.2.11. Workplan

A workplan outlines a set of planning activities to be undertaken during the preparation phases of implementation (for example, research activities), the sequence of activities, and individual responsibility for each activity. While the workplan may be revised later, it should set forth as precisely as possible what planning activities will be undertaken and by whom. The workplan should also indicate budgets and schedules for each activity (DENR *et al.*, 2001b) (Boxes 6.21 and 6.22).

Box 6.21. Key Elements of a Workplan.

Title
    Introduction
    Objectives
    Overview of tasks and outputs
        Task 1: Prepare programme
        Task 2: Identify stakeholders
        Task 3: Collect and analyse secondary information
        Task 4: Conduct research
        Task 5: Prioritize issues
    Staffing needs
    Schedule of activities
    Cost estimates
    Deliverables

Source: DENR *et al.* (2001b).

Box 6.22. Workplan for the Pilot Project on Co-management of the Sea Egg Fishery in Barbados.

Background:
The fisheries authority and fishing industry are interested in instituting community-based co-management, involving fishers in all aspects of management. This may include monitoring urchin size, maturity and population density, determining when and where fishing would be allowed and otherwise regulating the fishery to the extent that fisher knowledge and observations could be the main inputs to management. This pilot project will assist the stakeholders in pursuing their shared interest in co-management in a manner consistent with the Barbados 2001–2003 Fisheries Management Plan. The Fisheries Division and Barbados National Union of Fisherfolk Organizations are already collaborating on surveys at sea.

Objective:
The objective is for the fisheries authority and fishing industry to collaboratively determine and demonstrate the feasibility of co-management arrangements for the Barbados sea egg fishery within the period of the 2001–2003 Fisheries Management Plan.

Workplan:
• Collaborative surveys going from design to execution.
• Workshop on data analysis, generation and use of information as a demonstration of shared learning and to evaluate further development of their collaborative processes.
• Public education will be offered via a newspaper supplement, TV promotion, Fisherfolk's Week panel discussion, brochure, poster, radio or other methods, the effectiveness of which will be evaluated.

Outputs:
• Data from the sea urchin surveys in which fishers participate.
• Understanding about how information for management decision-making can be generated, shared and used in co-management.
• Increased public awareness about sea urchin management, leading to better compliance.

Source: McConney *et al.* (2003a).

7 Research and Participatory Research

J. Parks.

The next major activity in the co-management process is research.

The role of research in co-management is to help establish baselines, inform the management process, and nourish community education and involvement. A common mistake is to focus on research to the exclusion of education and action. By involving community members in these activities, the research process itself becomes one of education and action. In this way, such participatory research lays the foundation of awareness and commitment from which other activities grow.

(White *et al.*, 1994)

Table 7.1. Research and participatory research.

Stakeholder	Role
Resource users/community	• Attending meetings and briefings • Prepare workplan • Participate in research activities • Provide information • Willingness to learn new skills
Local government	• Attend courtesy calls • Participate in meetings and discussions • Assist in organizing community meetings • Assist in identification of community boundaries • Assist preparation of workplans
Other stakeholders	• Attending meetings and briefings • Participate in research activities • Provide information
Change agent/community organizer	• Organize courtesy calls to government leaders • Orient to situation • Organize community meetings • Observation • Identify stakeholders • Prepare workplan

Research constitutes the information gathering activities of the co-management programme (Table 7.1). A great deal of information is gathered about coastal resources, resource use activities and people. During this activity, both secondary and primary data are collected and analysed and a community profile is prepared. The community profile will serve as the basis for planning and management activities and as a baseline for future monitoring and evaluation. The community profile incorporates the community's problems, needs and opportunities assessment. The decision on the scope and scale of the community profile and research is made by the core group, based on information needs for decision-making and on available resources and time.

The community profile includes five components:

• Resource and ecological assessment;
• Socio-economic assessment;
• Legal and institutional assessment;
• Problems, needs and opportunities assessment;
• Management issues and opportunities.

While some of the information used in the community profile comes from secondary sources, other information will come from scientific studies by experts and from participatory research with resource users and other community members (Box 7.1). Scientific information is very useful and important for the community profile, but the type of information collected by scientists often differs from that obtained from resource users, and the tools and

Box 7.1. Joint Research in the Sokhulu Subsistence Mussel-harvesting Project in South Africa.

The Sokhulu mussel co-management project was initiated in 1995 to address problems of illegal harvesting and overfishing. One purpose of the project was to investigate the extent and impact of subsistence harvesting on the coast. One research activity was to investigate the use of three different harvest tools. The joint experiment to evaluate the efficiency and impact of different collecting tools laid the foundation for the modus operandi of decision-making. The exercise took place early in the project and demonstrated the concept of an experiment, the value of research and the principle of joint decision-making. It paved the way for the large-scale participative experiment to determine sustainable harvest levels. Community monitors participated in field surveys of stock abundance and helped to process samples. Their close interaction with the government project staff and familiarization with research techniques proved invaluable when explaining these matters to the harvesters.

Source: Harris *et al.* (2003).

methods of collecting the information are also different (Walters *et al.*, 1998). A significant amount of information can and should come from the community. The 'indigenous or local knowledge' of resource users and other community members (including women and elders) is critical information for planning and management. As will be discussed, there are a number of tools and methods available that involve the extensive participation of local community members in gathering and analysing information and for obtaining indigenous or local knowledge. The combination of scientific and local knowledge can complement each other and greatly enhance the co-management planning and management programme. The collection of information may take several weeks to several months depending upon the scope and scale of information needs. All the information collected should be kept in a small reference library so that it is available to all participants in the co-management programme.

Some of the data to be used in the community profile should have been collected during the preliminary community profile conducted in the community entry and integration phase (see Chapter 6, Section 6.2.4).

7.1. Participatory Research

The conventional approach to research has been characterized by control by outside experts, scientists and development specialists who have set project agendas and carried out research without any or only minor input from local community members (Chambers, 1994). Not only have local people not played a part in the planning and implementation of such projects, but their knowledge of local ecology and the structure of their social, economic and political systems have also been ignored. The process is relatively static, one in which information is gathered from a community and then processed and analysed by experts with little or no feedback to the community. Consequently, many projects have failed due to inappropriate project goals, community apathy and a lack of understanding of local social and ecological systems (Landon and Langill, 1998).

In recent years, new approaches to research have been developed which involve community members in gathering information in a participatory manner (Box 7.2). Participatory research (PR) represents a family of methodological approaches increasingly accepted and utilized to involve local people in research projects taking place in their own communities. PR is characterized by a cyclical, ongoing process of research, reflection and action, which seeks to include local people in designing the research, gathering information, analysing data and taking action. A key objective of PR is to empower community members by utilizing local knowledge and practices and by giving local people the opportunity to learn skills about and share in the research process. It is meant to move away from dependence on scientific information provided by outside professionals to local knowledge and skills. It is also intended to contribute directly to positive changes in the specific circumstances of the participants, as well as increase the chances that the co-management programme will succeed through local involvement (Landon and Langill, 1998).

There are a variety of ways that people can 'participate', depending upon the particular context of the research, the capacities of those involved and the willingness to let community people participate. Participation can range from consultation or information sharing (where local people are kept informed about research activities but do not influence the research process) to self-mobilization (where the researcher acts only in a guidance capacity and local people take the initiative in project design and implementation).

It is important to note that the use of participatory research does not mean that conventional scientific research conducted by outside experts should not

Box 7.2. Collaborative Surveys and Analysis in the Sea Egg Fishery of Barbados,

In developing a co-management pilot project for the sea egg (sea urchin) fishery in Barbados, one of the items identified in the workplan was collaborative surveys involving the fisheries authority and fishing industry going from design to execution. Fieldwork led by the Fisheries Division's fisheries biologist involved organizing fishers from around the island into four small teams covering 26 survey sites in specific segments of coastline. The biologist and assistant first explained the research design and methods in the classroom, followed by demonstration and practice in the field. Surveys were carried out by the fishers on several occasions. The 16 volunteer fishers brought their raw data to the biologist who collated it and conducted analyses. The analyses were explained to the fishers. The fishers entered data into the computer in order to get a hands-on feel for the mechanics of data processing. The information generated by this collaborative research was used to prepare a policy paper for the re-opening of the sea egg fishery.

Source: McConney *et al.* (2003a).

be conducted. There are a number of circumstances where this will be the most appropriate method of gathering information. There are overlaps and complementarities between the two approaches to research (Box 7.3). For example, participatory research may identify an issue, which is then studied in more depth through conventional research.

Conventional research is most appropriate:

1. When data needed are mostly quantitative;
2. When follow-up action, in terms of programme and project training, is uncertain;
3. When issues addressed are not sensitive;
4. When the purpose of the study does not include setting the stage for staff or community involvement in a programme;

Box 7.3. Conventional versus Participatory Research.

	Conventional research	Participatory research
Purpose	To collect information for diagnosis, planning and evaluation	To empower local people to initate action
Goals of approach	Predetermined, highly specified	Evolving, in flux
Approach	Objective, standardized, uniform approach, blueprint to test hypothesis, linear	Flexible, diverse, local adaptation, change encouraged, iterative
Modes of operation	Extractive, distance from subject, focus on information generation	Empower, participatory, focus on human growth
Focus of decision-making	External, centralized	Local people, with or without facilitator
Methods/techniques	Highly structured focus, precision of measurement, statistical analysis	Open-ended, visual interactive, sorting, scoring, ranking, drawing
Role of researcher/ facilitator	Controller, manipulator, expert, dominant, objective	Catalyst, facilitator, visible initially, later invisible
Role of local people	Sample, targets, respondent passive, reactive	Generators of knowledge, participants active, creative
Ownership of results	Results owned and controlled by outsiders, who may limit access by others	Results owned by local people, new knowledge resides in people
Output	Reports, publications, possible policy change	Enhanced local action and capacity, local learning, cumulative effect on policy change, results may not be recorded

Source: Narayan (1996).

5. When time and resources are not serious considerations.

Participatory research is most appropriate:

1. To establish rapport and a commitment to use study results;
2. When staff or community interest and involvement is central to achieving programme goals;
3. When information is sensitive;
4. When major issues are unknown or relatively undefined;
5. When supporting local capacity is important (Narayan, 1996).

Among the many approaches which make up participatory research, rapid rural appraisal (RRA) and participatory rural appraisal (PRA) offer the best alternatives to the conventional approach to research. Although there are a large number of descriptions, the most useful are the following.

Rapid rural appraisal (RRA) emphasizes the importance of learning rapidly and directly from people. RRA involves tapping local knowledge and gaining information and insight from local people using a range of interactive tools and methods (Jackson and Ingles, 1995). RRA is often used to inform those from outside the community (Box 7.4).

Box 7.4. Rapid Rural Appraisal.

Rapid rural appraisal (RRA) consists of the following features:

- An activity carried out by a group of people from different backgrounds (usually from outside the community or area), which usually aims to learn about a particular topic, area, situation, group of people or whatever else is of concern to those organizing the RRA;
- It usually involves collecting information by talking to people in the community;
- It uses a set of guidelines on how to approach the collection of information, learning from the information and the involvement of local people in its interpretation and presentation;
- It uses a set of tools that consist of exercises and techniques for collecting information, means of organizing that information so that it is easily understood by a wide range of people, techniques for stimulating interaction with community members and methods for quickly analysing and reporting findings and suggesting appropriate action.

    RRA guidelines include:

- Structured but flexible: clear planning and objectives but flexible to respond to changing conditions and circumstances.
- Integrated and interdisciplinary: a team composed of people from different disciplinary and skill backgrounds.
- Awareness of bias: team is aware of their own and respondents' bias and cross-check results.
- Accelerating the planning process: tries to shorten the time to know an area and plan for interventions.
- Interaction with and learning from local people: must involve local people who are the intended beneficiaries of the results.
- Combination of different tools: combination of communication and learning tools.

- Iterative: constant review of results.

   The RRA toolbox is broad, varied and constantly growing. The tool is chosen for a specific objective, context and conditions and should reflect 'personal' tools of the user. In a broad categorization, RRA tools include:

- Secondary data review;
- Workshop;
- Structured observation;
- Ranking and classification;
- Interviews;
- Community meetings;
- Mapping techniques;
- Diagrams and graphics;
- Understanding processes and change.

   A typical sequence of RRA activities includes:

- RRA preparation (setting objectives, team identification, secondary data);
- Preliminary workshop of team and other concerned people (training of team, review of secondary data, choosing tools, planning);
- First fieldwork session;
- Intermediate workshop (team review of findings, revision of objectives);
- Second fieldwork session;
- Intermediate workshop (team review of findings, revision of objectives);
- Third fieldwork session;
- Intermediate workshop (team review of findings, preparation of draft report, preparation for community meeting);
- Community meeting (presentation of findings, discussion and correction of findings, definition of future action);
- Final workshop (final review of findings, final report, future action plans).

There are four broad categories of RRA:

1. Exploratory: purpose of learning about conditions in a particular area with a view to designing appropriate development activities.
2. Topical: purpose is to learn more about specific issues in order to understand them more completely.
3. Monitoring and evaluation: purpose is to monitor and evaluate ongoing activities.
4. Participatory: purpose is to have more community member involvement in the approach.

Source: Townsley (1996).

   Participatory rural appraisal (PRA) involves field workers learning with local people with the aim of facilitating local capacity to analyse, plan, resolve conflicts, take action and monitor and evaluate according to a local agenda (Jackson and Ingles, 1995; Maine *et al.*, 1996) (Box 7.5).
   RRA is regarded as a set of guidelines and tools which can be used in many different ways and many different circumstances and without necessarily attempting to change political and social structures. PRA is used to specifically

> **Box 7.5.** Participatory Rural Appraisal.
>
> The distinction between RRA and PRA can sometimes be considered as academic since both include scope for the participation of local people in the activity. PRA is not just a tool which enables development planners to learn about rural conditions and consult with local people so that they (the development planners) can come up with more appropriate and better development plans. Instead, PRA is sometimes regarded as an exercise which transfers the role of planning and decision-making, traditionally taken by the government and development agencies, to the target group or community itself.
>
> Unlike the RRA, the PRA responds to the needs of the community and target group, not the development workers. In PRA, the tools are used to help people analyse their own conditions and communicate with outsiders, rather than in an RRA where tools are used by outsiders to understand local conditions. Focus of the PRA is decided by the community and it empowers them to communicate their needs to government and development agencies.
>
> Source: Townsley (1996).

refer to a use of RRA approaches and tools to encourage participation in decision-making and planning by people who are usually excluded.

Both approaches are carried out by multidisciplinary teams and differ from conventional information-gathering approaches in that field workers work with and learn directly from local people. The methods involve a minimum of outsider interference or involvement.

Because of the apparent simplicity of the methods, many feel that they can be learned from a book or in the classroom. However, the subtleties of understanding and using the methods can be learned only by experience. RRA is a useful technique for data gathering and problem identification, whereas PRA is more appropriate to programme design and planning. The distinction is not merely one of proper sequencing. If not used correctly, PRA can generate false expectations of what the programme will provide or what local people can achieve. This can cause problems in the relationship between the community members and the programme staff which can threaten success.

While the speed with which data can be collected using these methods is an advantage, it also introduces the risk of accepting superficial descriptions of situations and adopting superficial solutions to issues. RRA and PRA are often seen as a replacement for other forms of investigation and study even in situations where more formal or analytical research is called for. These methods can be biased towards local people who have time and motivation to talk to field workers and towards people who appear to have knowledge. The potential effectiveness of the RRA and PRA methods can be reduced by either using them in a too highly formalized fashion, or applying them too rigidly and repetitiously. Good practitioners will use a variety of methods for cross-checking information and the applicability of proposed solutions (Box 7.6). The depth of understanding of issues should be undertaken over time and with other methods (Box 7.7).

Box 7.6. Guidelines for Participatory Research.

- Set objectives first so that the most appropriate tools can be selected.
- Build on previous information gathered. The results of each tool can be used to generate new ideas.
- Cross-check and probe to ensure reliability of information.
- Analyse and validate on the spot.
- Avoid collecting information that is not necessary.
- Avoid bias.
- Listen to the community leaders but recognize that they may be the local elites and have their own bias.
- Acknowledge the value of indigenous knowledge, skills and practice.
- Be creative.

Source: IIRR (1998).

Box 7.7. Research for Management of Tam Giang Lagoon in Hue Province, Central Vietnam.

One of the issues in the Tam Giang lagoon is the narrowing of waterways because of the increase in the number of net enclosures in the lagoon. Waterways are the areas between fish corrals and bottom nets where fishing activities are not allowed and managed as common property. Net enclosures have been set up with the approval of local governments through taxes imposed on aquaculture production. The resulting narrow waterways have caused navigational accidents, particularly among mobile gear fishers. The fry migrating from the sea have also been affected and water exchanged has decreased.

The researchers of Hue University served as links between the government and the community people in attempting to solve this problem. They attempted to create an atmosphere of cooperation that is based on the principles of honesty, faith, respect, trust, effectiveness and interest. A Government–Fisher Joint Committee for Research was formed to lead the investigation of the problem on the narrow waterways and recommend workable solutions that will be acceptable to all parties concerned. With this process, the fishers felt they were making a contribution to solving their problems. The local government officials, on the other hand, deepened their understanding of the lagoon ecology and became increasingly aware of the potential benefits of resource management. What was important in this process was having a common goal that revolved around the resources of the lagoon, the livelihoods of fishers, fishing and aquaculture activities and the local fishery management system. It was important too that sustained environmental awareness activities were implemented and constant dialogues and discussions were held. The researchers became environmental educators, negotiators and intermediaries, especially among the people involved in aquaculture who resisted the most in fear of losing their fishing ground.

The research results led to the implementation of a pilot project and based on agreed criteria, the Xa Bac waterway was selected. To establish and manage the project, a joint committee for development and another for management were set up. The members of the committees were selected by the fishers. The committees set up the rules in a participatory manner and these were discussed and evaluated in a workshop they called 'Nursing Co-management in Phu Tan Area'. Both government and fishers acknowledged the usefulness of the cooperative approach to solving the problem in the waterway. They both acknowledged the lesson of listening to each other and combining their strengths to solve problems. The challenge now is to build on this positive experience and replicate it in the adjoining areas in the lagoon.

Source: Phap, Tôn Thât (2002).

It should be noted that most community members have limited experience with participation and participatory research methods and time will need to be allocated to empower community members to actively participate in PRA activities. In addition, PRA should not be thought of as merely about tools that would lead to some initial form of community analysis about their problems and issues. PRA, more than anything else, is about the shared experience of the researchers and communities in a cycle of reflection and action. It is not just doing research with local people; it is part of a much broader goal of people empowerment (Brzeski *et al.*, 2001).

Good RRA and PRA are characterized by behaviour and attitudes that build rapport with local people, avoiding putting people in uncomfortable situations, learning from people not lecturing to them, creativity, checking and rechecking the validity of the information obtained, listening and probing, being patient, engaging in conversations that have two-way exchange of information, being trustworthy, and being open and friendly (Jackson and Ingles, 1995).

7.2. Indigenous Knowledge

Indigenous knowledge (IK) is broadly the knowledge used by local people to make a living in a particular environment. Terms used also include traditional environmental or ecological knowledge, rural knowledge, local knowledge and fisher's knowledge. IK can be defined more specifically as: 'A body of knowledge built up by a group of people through generations of living in close contact with nature' (Johnson, 1992). Generally speaking, such knowledge evolves in the local environment, so that it is specifically adapted to the requirements of local people and conditions. It is also creative and experimental, constantly incorporating outside influences and inside innovations to meet new conditions. It is usually a mistake to think of IK as 'old-fashioned', 'backwards' or 'static'.

IK is considered to be both technical knowledge of the environment, as well as cultural knowledge including all of the social, political, economic and spiritual aspects of a local way of life. IK includes classification systems of fish, animals, terrestrial and aquatic plants, soil, water, air, weather; empirical knowledge about flora, fauna and inanimate resources and their practical uses; resource management knowledge and the tools, techniques, practices and rules related to fishing, gathering of wild food, agriculture and agroforestry; and the world view or way the local group perceives its relationship to the natural world. Such knowledge can contribute to resource management, development of alternative economic strategies, conservation, environmental assessment and biological and ecological research (Box 7.8).

IK can provide valuable input about the local environment and how to effectively manage its natural resources. Also, by incorporating IK into projects it can contribute to local empowerment, increasing self-sufficiency and strengthening self-determination. Its use by outsiders to the community can

Box 7.8. Fisheries Assessments: What can be Learned from Interviewing Resource Users?

The body of information held by fishers has an important role to play in fisheries assessment. When this body of information matches scientific assessments, uncertainty is reduced and assessments become more convincing to resource users. When the two sources of information diverge, information from both sources needs to be re-examined. Yet while this body of information has a role to play, there remain practical impediments to its use. Scientific terms do not match the terms that fishers use to organize their knowledge. The geographic range of information from each fisher is limited. Knowledge is unevenly distributed among fishers, being more concentrated among older fishers and skippers. It is largely oral, rather than written, and subject to the effects of memory loss.

Interviews with fishers can be used to obtain large amounts of information on fish behaviour and fishing patterns. Local knowledge of the dates when fish are caught in fixed-gear location can provide information on seasonal and directional fish movements. Fishers can provide information pertaining to stock structure. Fishers can provide information on movement patterns (through catch patterns), spawning grounds (presence of female's ripe and running condition), juvenile habitat, and spatial patterns in fish morphology. This information is useful in conjunction with genetic information, tagging experiments and morphometric studies used in identifying stocks. Catch rate data obtained from fishers have the potential to reflect local changes in fish abundance.

Source: Neis *et al.* (1999).

increase cultural pride and thus motivation to solve local problems with local ingenuity and resources (Boxes 7.9, 7.10 and 7.11).

As with scientific knowledge, however, IK has its limitations, and these must be recognized. IK is sometimes accepted uncritically because of naïve notions that whatever indigenous people do is naturally in harmony with the environment. There is historical and contemporary evidence that indigenous

Box 7.9. The Uses of Indigenous Knowledge in Marovo Lagoon, Solomon Islands.

As in other strongly maritime-oriented Pacific Island societies, an extensive body of environmental knowledge underpins the impressive array of Marovo fishing methods. Decisions on fishing are made based on this knowledge, while taking into account the constraints of marine tenure regulations, which indicate what fishing grounds and fishing technologies may be legitimately used. The local classification of fish habitats includes more than 40 terms for district reef features, water depths and bottom types. The migration paths of crabs, crayfish and molluscs are known. There are gender differences in these fields of indigenous knowledge. Men, for instance, pride themselves on understanding fish spawning behaviour to the extent that, for many food species, they can accurately predict its occurrence. Women hold extensive knowledge of daily, lunar and seasonal rhythms in the abundance and distribution of molluscs and crustaceans.

Source: Hviding and Baines (1996).

Box 7.10. Mataw Fishing in Batanes, Philippines.

Mataw fishing involves the traditional capture of seasonal flying fish in Batanes Island, Philippines. Mataw fishermen are organized as associations of users of vanua, the natural access ways for a boat allowing transit between land and sea. The Mataw associations have their own economic arrangements, observe local laws and perform rituals for the vanua. Without any external support, the Mataw associations have locally negotiated the uncertainties of fishing, the hazardous environment personified by invisible spirit beings, and the competition from fellow fishers, through the observance of taboos, rituals and laws. For example, the rights to fish and use the vanua safely are gained by conducting an exchange through ritual sacrifice with the anitu or invisible spirit beings. The vanua becomes a sacred area for the duration of the fishing season. Mataw organizations regulate access and exploitation of resources within the vanua and traditional fishing grounds, under the leadership of the ideal fisherman who makes the first fishing trip for the season and who possesses the power to ritually set precedents for the season. With the present innovations in technology and other historical trends, mataws, both as individuals and as members of associations, are seen to creatively negotiate the conflicting interests of fellow fishermen in the face of the opposed values of the indigenous world view and dominant modernizing paradigm.

Source: Mangahas (1993).

peoples have also committed environmental mistakes through over-fishing, over-grazing and over-hunting.

A critical assumption of IK approaches, for example, is that local people have a good understanding of the natural resource base because they have lived in the same, or similar, environment for many generations, and have accumulated and passed on knowledge of the natural resources and

Box 7.11. Using Indigenous Knowledge.

The use of IK in a programme can be evaluated through a five-step process:

1. Identify the problem or issue for which information is sought.
2. Working together with community members, record and briefly document all IK available in the community relating to the problem, including what has been done in the past and what is done now. If no IK exists, it might be necessary to test, adapt and promote scientific knowledge.
3. If relevant IK does exist, local people and field workers can together discuss and screen their findings, looking for the most relevant IK information. Understand the reasons behind a particular practice or belief.
4. Test whether the IK can be improved. It may be possible to blend IK and scientific knowledge.
5. The improved IK can be promoted through information exchange and extension.

Source: IIRR (1998).

conditions. In some cases, local people may be recent migrants from other areas and may not have accumulated much indigenous knowledge about their new environment. It is important to evaluate the relevance of different kinds of indigenous knowledge to local conditions. Most observers, in fact, suggest that a combination of both IK and science be used.

In order to utilize IK, it must first be documented, and researchers must be aware of the ethical and methodological issues associated with doing research in local communities. The methods of participatory rural appraisal, for example, are now accepted as a means of effectively involving local people in the research process. Researchers must also be sensitive to the issue of intellectual property rights over knowledge. Local people have become concerned that knowledge is being 'stolen' and used without their awareness and without a share in any economic benefits that may result from the development of related commercial products. This knowledge must, therefore, be protected (Langill and Landon, 1998). The ethics of IK also require that those with and who provide the knowledge be asked first whether they will share that knowledge and be acknowledged for doing so.

Older people have different types of knowledge than the young. Women and men, net fishers and spear fishers, educated and uneducated people all have different types of knowledge. Common knowledge is held by all people in a community (e.g. how to cook rice). Shared knowledge is held by many but not all community members (e.g. seasonal fishing activities). Specialized knowledge is held by a few people who might have served an apprenticeship (e.g. location of fish aggregations). The types of knowledge held by people is related to their: age; sex; education; labour division in the family, enterprise or community; occupation; environment; socio-economic status; experience and history.

7.3. Gender

Participatory research requires full community participation, including women, men, youth and elderly. Gender differences are often easily ignored given that most fishers who go to sea are men and development projects tend to be male-dominated. Equitable impact of development is enhanced by putting views, areas of knowledge and strategies of women into the process. Understanding these issues can have important implications for project planning and to reduce conflict (Box 7.12).

Women play a valuable role in the many activities in the community and can provide much useful information on such issues as time use, access to resources, food distribution, decision-making and control of income. In fisheries, for example, women often are involved in gleaning and other shore-based harvesting activities. Women also are commonly fish traders. Efforts should be made to encourage women to participate in PR activities. In some cultures, it may be socially uncomfortable for women to participate. In these cases, accommodations should be made, such as having separate exercises for men and women, or using a woman facilitator.

> **Box 7.12.** Gender-sensitizing Efforts in the Philippines.
>
> A Philippine NGO, known as Community Extension and Research for Development (CERD), has institutionalized gender programmes in its work in the fisheries sector. In 1991, a member of CERD was involved in a participatory research project called Gender Needs Assessment (GNA) of women in the fisheries sector. Findings of the research showed the major roles women play in fisheries and reaffirmed the earlier findings of CERD research. The staff involved in the project re-echoed the training to other CERD staff and they formed a team that piloted GNA in selected programme areas. The GNA facilitated the reorganization of a women's organization in one village and this led to the formation of another organization in a neighbouring village. The women who participated in the GNA realized that they had the potential to solve their problems and that they could have the strength to do this if they unite and organize themselves just like the men in their community.
>
> Source: Cleofe (1999).

Understanding the roles women and men play in the household and in the community as they relate with resource use and management is the first step in analysing gender. However, it is not enough to end here because a deeper analysis on gender relations that includes an understanding of the dynamics of men–women relationships and patterns of decision-making is needed.

7.4. Community Profile

The community profile will serve as the basis for planning activities, provides a context for management and serves as the baseline for monitoring and evaluation. It provides the community with information about itself, often information which the community may not even know or be aware of itself, such as similarities and differences and views and attitudes. The community profile should be sufficiently detailed to provide the reader with a clear understanding of the environmental and social conditions at the site, why management is needed and how management might improve coastal conditions (DENR *et al.*, 2001c). The profile should help to answer key questions:

- What are current resource conditions, patterns of resource use, and resource use problems and how are they changing over time?
- What problems or obstacles for coastal management are revealed?
- What are patterns of power as they relate with resource use and exploitation? And gender differences?

From a CO perspective, it is also important for the community profile to contain community responses to problems and issues because this gives us an idea of what issues resulted in some collective action, and in what ways the community members cooperated with one another.

A community profile should contain detailed data on aquatic habitat distribution, resource conditions, demographic and socio-economic conditions, existing legal and institutional arrangements for resource management, and identification of problems, needs and opportunities. Detailed maps are included to better illustrate the habitats, resources and socio-economic activities.

As mentioned above, the community profile includes five components:

- Resource and ecological assessment (REA);
- Socio-economic assessment (SEA);
- Legal and institutional assessment (LIA);
- Problems, needs and opportunity assessment (PNOA);
- Management issues and opportunities.

Each component of the community profile will involve use of different methods, but all combine participatory research and scientific assessment.

The preparation of the community profile is a multi-step process for gathering information:

1. Preparation;
2. Secondary data collection;
3. Fieldwork/assessment/research;
4. Database and profile preparation;
5. Prioritizing results and analysing causes;
6. Validation.

7.4.1. Preparation

Preparation is when the scope of the community profile is established. This involves:

- Team organization;
- Defining the goals and objectives;
- Identifying the stakeholders;
- Determining the parameters to be included.

The community profile addresses a broad range of issues across different disciplines and technical fields, including social sciences, natural sciences and political sciences. Ideally the team conducting the community profile, working with community members, will reflect this range by including social scientists, natural scientists and political scientists. The most important consideration for the team is a balance of expertise. It may be feasible for a team, at a minimum, to be composed of two technical experts (for example, a marine biologist and an economist) complemented by local researchers. Local researchers, coming from the community or an NGO, bring in local knowledge and their skill base may be expanded by participation in the community profile activities. An ideal community profile team may be composed of a coastal habitat expert, fisheries biologist, economist, sociologist or anthropologist, and political scientist or public policy expert. The team leader will be critical to guide and organize (logistics, administrative, community relations) the community profile. The team leader may be a technical expert or local researcher or may have no technical assignment and only handle leadership activities. Team members may be divided into sub-groups based on component, for example REA, SEA or LIA. Team members may be rotated to conduct field work.

In addition to their disciplinary and technical skills, team members should also have the following characteristics:

- Open-minded attitude and willingness to learn;
- Gender balance;
- Ethnic balance;
- Local language skills;
- Organizational background.

A challenging aspect in combining expertise is trying to make individuals work together. For example, an initial activity is team building so that the sociologist will have a chance to know the biologist better, and vice versa. Social scientists can learn how to dive. In the field, even if the division of tasks is quite clear-cut, there should be regular sharing sessions on field experience and findings.

The goal of the community profile will guide how to set up the activities and how complex it will be. In most cases, the goal will be to increase knowledge of the biophysical, socio-economic and institutional conditions of the community for planning and management. The goal may also include monitoring and evaluation in the future so there will be a need to establish a good baseline for assessing change over time. Specific objectives are defined to identify more specifically the focus and activities of the research.

The process of conducting the community profile will need to be defined including determination of time, people, funding and resources needed for each component. A table can be drafted which identifies each component and activity, timeline, resource needs and team members.

The team will need to identify the community members and stakeholder groups to determine which ones should be the focus of the activities. Depending upon the goals of the community profile, the study may only involve those stakeholders associated with particular activities. When it is not possible to study all stakeholders, it may be necessary to set priorities for which stakeholders to study. This can be done by noting three main factors:

- Their proximity to the resources;
- The impact that their activities may be having on the resources;
- Their relative levels of dependence on resource-related activities.

New stakeholders may be identified as the team learns more about the area.

Stakeholders who will be the focus of the community profile activities should be consulted early to help ensure concerns and priorities are addressed, ensure cooperation, increase stakeholders' sense of ownership of the study, involve them in the study and increase support for the output. Consultations can be conducted through one-on-one meetings, small discussion groups and community meetings.

The level of participation of stakeholders may change during the research. The team will need to decide on the right level of participation, level of interest, which research activities stakeholders can be involved in, the resources (number of people needed, training, funds) required and the political context for participation.

A rule of thumb for most researchers is to encourage as much participation as possible from all stakeholders. However, the determining factors would be which stakeholders are appropriate to include in relation to the research objectives and the willingness of stakeholders to participate. Participation is not something only the researchers decide on (this immediately violates the essence of PR). The researcher strives to have community members (i.e. stakeholders) as equal partners in research and would always seek the broadest and deepest participation possible from all stakeholders.

The team needs to decide which parameters and sub-parameters to research for each component. The parameters determine the substance of the research and form the basis for deciding which methods will be used in the field. There is no definitive list of parameters for the REA, SEA and LIA. The sections discussing each of these components below will provide a listing of the most commonly identified parameters. Rarely is it possible to assess all the parameters. Therefore the team will need to prioritize based on goals and objectives, needs of the end-users, and resources and time available.

7.4.2. Secondary data

All relevant secondary data on the parameters and sub-parameters are identified. Secondary data are those that have already been collected, analysed and published in various forms, including:

- Official and unofficial documents;
- Statistical reports;
- Reports of previous assessments and surveys;
- Research reports, including academic papers, e.g. thesis;
- Documentation of previous or ongoing projects, including monitoring and evaluation reports;
- Maps;
- Aerial photographs and satellite images;
- Historical documents and accounts;
- Websites on the internet.

These data will be used to:

- Identify gaps in existing knowledge in preparation for the field data collection;
- Ensure the field data collection does not collect information that has already been collected;

- Provide a basis for cross-checking information collected during the field data collection;
- Provide supporting documentation for the field data collection;
- Refine the lists of objectives, stakeholder groups, study sites and parameters.

Some secondary data should have been collected during the 'beginnings' stage. The team will want to review this and decide if additional secondary data are needed and are available.

In the process of collecting secondary data, the team may make contact with individuals who have worked in the community or area in the past. In addition to the materials provided, these contacts can often provide valuable information and insights from personal experience.

The search for secondary data should go beyond the usual government offices. Unofficial materials or 'grey publications' exist from student theses and NGO project activities. There is no detailed methodology for gathering secondary data. It is mostly a matter of writing letters, making telephone calls, visiting offices and libraries, interviewing officials, teachers, scientists and researchers. When a document is located, check the references section and try to locate relevant references. Always credit the source of the material in the community profile.

The secondary data should be compiled so that it can be easily used during the research. A filing system should be developed to code, record and store the secondary data. The team should read through the secondary data to identify information related to the parameters and to determine the quality of the data.

7.4.3. Reconnaissance survey

A reconnaissance survey, which is a brief observational survey of the study area, can provide the team with valuable information to help plan the field data collection. This activity allows the team members to familiarize themselves with important field features, such as terrain, natural resources and human settlement, and to know firsthand the feasibility of conducting the research. It can also help identify logistical requirements based on local conditions and make arrangements for the field data collection. Where the team is familiar with the area, a brief trip may be needed. If the team is unfamiliar with the area, the reconnaissance may involve collecting some basic preliminary data through a rapid survey of stakeholders and resources. Key informants may be interviewed and biophysical characteristics mapped.

7.4.4. Planning of the field data collection

The team should plan the field data collection in detail to ensure that they will enter the study area prepared to collect the data effectively and efficiently. Planning the field data collection involves several steps:

- Identifying the methods for each component and the parameters;
- Preparing materials and tools for the methods;
- Pre-testing instruments, equipment, interview guides and questionnaires;
- Deciding how to keep track of information;
- Develop a database that allows for analysis and information retrieval;
- Developing a coding system for data;
- Defining plans for analysis;
- Deciding on sampling units;
- Deciding who to interview and survey;
- Establishing the field teams;
- Defining the schedule for field data collection;
- Training team members;
- Providing a briefing on local culture;
- Arranging logistics.

The entire team needs to be involved in planning. This develops team spirit and ensures everyone understands everything involved in the field data collection. Regular team workshops should be held throughout the field data collection to exchange information and identify problems and needs.

7.4.5. Resource and ecological assessment

Resource and ecological assessments (REAs) are detailed studies which may include biological and physicochemical parameters. The information obtained can be used to determine the status of the ecosystem. The reports or profiles generated are technical and quantitative in nature. REAs are usually conducted by highly skilled technical persons mostly coming from academe. Methodologies are generally based on English *et al.* (1994).

REAs can also be conducted with community participation with very minimal technical input. When outside experts conduct the REA they can benefit from participation by local community members and can provide them with training in methods which can be used for monitoring studies. Time should be taken to explain the characteristics, terminologies, uses and analysis of the methods to the local people. Usually resource maps, transects and trend diagrams are generated as a result of consensus among community participants (Walters *et al.*, 1998) (Box 7.13). This approach increases community participation and can be conducted at a minimal cost. Maps and data produced in a participatory way can be validated by more scientific assessments conducted by experts. In addition, time should be taken to listen to the local people discuss the coastal area and their use of the resources (Boxes 7.14 and 7.15).

An example of community participation in a REA is a commercial fish landing survey. This is a tool to assess commercially important fishery resources at the local level. Selecting appropriate fish landing sites and designing the questionnaire or survey form is the first step and will depend on the purpose of the survey and the type of information needed. Surveyors can

Box 7.13. Transects.

Transects are both a way of representing information and a technique for familiarizing with the different parts of the community and ecological zones in the area. Among the advantages of the transect is the simple portrayal of the resources present and the associated economic, social and environmental issues in spatial terms. Transects can be made by community members walking a transect from the sea to upland areas.
The transect is conducted by:

- Clearly identifying the information needs and preparing a workplan;
- Choosing the area direction and length of the transect;
- Identifying a transect or reference line;
- Assembling equipment;
- Choosing the time;
- Taking notes of observations made to the left and right sides every 50 metres. Explore the leftward and rightward areas from the reference line but always return to the line and resume the original path;
- Record significant ecological and resource use changes;
- At the end of the transect, return to the village to consolidate and cross-check the information;
- Use the information to draw a coastal profile.

Source: IIRR (1998).

Box 7.14. From the Fisher's Memory: Reconstructing the History of Catch Per Unit Effort and Finding Historical Levels of Biomass in Danao Bay, Philippines.

In 1997 and 1998, several gear users were invited to workshops. The participants were grouped according to the year they started to use the gear (usually by decade). They were then asked to recall when they started to use the specific technology and to answer the following questions:

1. What was your normal or average catch in the year you started to use the gear?
2. At that time, how many other fishers were using the same gear in Danao Bay?
3. What were the other gears used then in the Bay (both legal and illegal)?

Each group reported their findings, starting with the first (often the older) gear users. The results showed the steady decline in catch per unit effort, the increase in number of gear users and sometimes an increase in effort per gear user. Most of the time, a rather gloomy picture emerged, especially if the declining line was extended into the future. This was an essential part of problem analysis.

Source: Heinen (2003, p. 51).

Box 7.15. Fishers' Knowledge of a Newfoundland, Canada, Fishery.

Personal interviews were conducted with fishers in the Bonavista and Trinity Bay, Newfoundland area inshore small-vessel (<35 ft) and nearshore larger-vessel (>35 ft) fisheries. Three kinds of interviews were conducted: one to define terms, one that allows geographically limited information collected by individuals during their careers to be assembled to identify recurrent patterns of change, and one that allows verification, refinement and updating of information on particular fisheries.

Personal interviews were conducted with the fishers and follow-up telephone interviews were conducted several months later. Effectively assessing fishers' knowledge requires shared understanding of local terms for fish and fishing grounds and for fishing gear; thus, the study began with ten taxonomy/toponomy interviews. Interviews were conducted with a sample of fishers. Follow-up telephone interviews with a sub-sample of the fishers were used to supplement data on the lumpfish roe fishery collected in the first set of interviews. Results from the research were presented at local feedback meetings attended by some of the study participants and some who did not participate in the study.

Source: Neis *et al.* (1999).

go to the landing sites and record types, number and size of fish being landed, as well as number of boats, fishers and types of gear being used. Additionally, using interviews, information on changes in fish catch over time can be gathered by asking fishers what they are catching now, what they caught 5 years ago and 10 years ago. The community participants and the experts analyse the data together.

The REA content can include information on the following parameters:

- Physical setting (geophysical overview including: land; soil; slope; sea floor; coastal habitat classifications; overview of coastal forests, rivers and watershed);
- Ambient environment (salinity, turbidity, light penetration);
- Climate (seasons, rainfall, winds, temperature, cloud cover);
- Oceanography (bathymetry, current/circulation patterns, tidal flow, waves, water quality, eddies, runoff patterns, substrate);
- Important habitats (coral reefs, seagrass beds, mangroves, wetlands, beaches, soft-bottom, estuaries, lagoons and bays);
- Fish, crustaceans, molluscs, echinoderms, elasmobranchs, porifera, aquatic plants, marine mammals, seabirds and other aquatic life;
- Resource use (terrestrial and marine uses);
- Technical attributes of the fishery (type (artisanal, small-scale, commercial, industrial), gear/fishing technology, species harvested, level of exploitation);
- History of resource use (number of resource users, gear, catch, habitat).

Mapping is one of the most important REA activities. Mapping can be more accurately accomplished or verified with global positioning system technology (Walters *et al.*, 1998). Several types of maps can be produced:

- A sketch map is a freehand drawing that can reveal much about both coastal features and the people who prepare them. These maps start on a blank piece of paper and stakeholders identify major features and distinct features.
- A thematic map displays selected information relating to a specific theme, such as land use, coastal habitats, slope, elevation and soil. These may be qualitative (e.g. land use) or quantitative (e.g. population density). The requisite thematic maps are: land use; coastal habitats; resources; uses, livelihood and opportunities; problems and issues; and transect/cross-section.
- A base map shows selected features such as coastline, roads and villages and serves to orient the stakeholder and assist in accurately identifying features.
- A land use map refers to actual land cover or any form of man's use of land. Land use should also be viewed from an historical perspective.
- A spot map describes the area in terms of important features such as roads, rivers and cultural landmarks.
- A coastal habitat map shows the location of important habitats.
- A resource map is a summary of the spatial distribution and condition of resources in the area, and includes resources that provide food and other materials of value to the community. Use, livelihood and opportunities include sites (e.g. fishing gear areas, gleaning sites, mining areas, mangrove cutting) where activities are accomplished or where opportunities or functions provide potential benefits to communities. Problems, issues and conflicts are mapped for later use in management.

A transect is a general reference line cutting across a representative portion of the study area. In effect, the transect line is the side view or cross-section of the site.

Before undertaking a REA, it is important to recognize that there is considerable natural variation within marine ecosystems, both spatially and temporally. Knowledge of the biology of the animals and plants being surveyed will help the researcher understand the differences observed in an ecosystem at different sampling scales. In order to accurately describe the communities in an ecosystem, survey programmes should be designed to minimize differences caused by the sampling itself. It is necessary to conduct regular monitoring to detect changes and suggest causes of change in a resource over time (English *et al.*, 1994). In undertaking the REA it is important that there is a common understanding of local names and terms among the scientists and local people. This is where folk taxonomy, undertaken as part of the SEA, becomes important. It may be useful to refer to photographs, pictures and actual samples or specimens to find out what local names correspond to scientific names. Fishers can be asked to identify habitats, resources, uses, issues and other features on maps.

A variety of detailed technical surveys are conducted in a REA. A few examples of these technical surveys include:

- *Manta tows:* manta tows involve visual assessment of large areas of underwater habitats by towing an observer behind a small boat. This technique is useful in assessing large-scale changes in resource conditions, determining the effects of disturbances on the underwater community, or in selecting sites that are representative of quality habitat for marine reserve status.

- *Line intercept transects:* line intercept transects are used to assess and describe the benthic community in coral reef habitats. Divers swim along transect lines placed along the bottom and record the percentage cover of life forms (rather than species) of major groups of corals, sponges, algae and other organisms. This is a reliable and efficient method of obtaining per cent cover data and spatial patterns in abundance of important groups of organisms.
- *Transect line plot:* the transect line plot is used in mangroves to determine the relative frequency, density and species diversity of mangroves. For each site, transect lines are drawn from the seaward margin of the forest at right angles to the edges of the mangrove forest. Plots are established at 10-metre intervals along a transect through the mangrove forest in each of the main forest types or zones. The method provides quantitative descriptions of the species composition, community structure and plant biomass of mangrove forests.
- *Visual fish counts:* visual census of fish abundance is an efficient and quantitative tool to evaluate fish abundance and diversity. A diver swims along transects laid on the bottom and counts fish observed within specified distances from the line. The type of fish counted can include all mobile species, target species for fisheries or indicator species.
- *Record books:* record books can be used by fishers to gather data on catch by gear type and fishing area. At a minimum, total catch per gear per fishing trip should be recorded.

7.4.6. Socio-economic assessment

A socio-economic assessment (SEA) is a way to learn about the social, cultural, economic and political conditions of individuals, households, groups, communities and organizations. There is no fixed list of topics that are examined in a SEA, however, the most commonly identified topics are: resource use patterns, stakeholder characteristics, gender issues, stakeholder perceptions, organization and resource governance, traditional knowledge, community services and facilities, market attributes for extractive use, market attributes for non-extractive use, and non-market and non-use values. SEAs vary in the extent that they cover these topics, and this will depend on the purpose of the assessment. Some SEAs may be a full evaluation of all these topics; others may focus on stakeholder perceptions or resource use patterns. Methodologies can be found in Bunce *et al.* (2000).

SEAs can be participatory (a broad range of people are involved in data collection, analysis and use) or extractive (outsiders conduct the assessment and take the information with them). They can also be product-oriented (report produced for a specific stakeholder group) or process-oriented (the process of collecting information is as important as the information).

SEAs involve planning and preparation before the assessment team (composed of experts, community members and programme staff) interacts with stakeholder groups through interviews and observation to collect field data. The assessment concludes with the team analysing and presenting the data. However, there is no best step-by-step way to conduct a SEA, and the order of the steps will vary widely depending upon local conditions and the requirements of the people. Sometimes, the assessment steps may follow a clear order, but in other cases they may need to be repeated and the order changed to adapt to new learning and changing circumstances. Each SEA should be adapted and the process modified to the situation faced by the team and based on experience, common sense and knowledge of the area.

There is a wide range of socio-economic parameters which can be included in a SEA. These include:

- *Resource use patterns:* the ways in which people use or affect coastal and aquatic resources such as resource-related use activities, stakeholders, techniques for resource-related activities, use rights, location of resource-related activities and stakeholders, and timing and seasonality of activities.
- *Stakeholder characteristics:* demographic characteristics of stakeholders including inhabitants and households, residency status, ethnicity, caste, religion, age, gender, education, family size, years of fishing experience, social status, household economic status, nutrition/health status, stakeholder livelihoods, household assets, community livelihoods, income sources, land tenurial status and culture.
- *Economic and political power relations:* each community has different levels of social and economic relations due to the economic and political power structures which exist both in and outside the community (Box 7.16).

Box 7.16. Economic and Political Power Analysis.

Every community has different levels of social and economic relations due to the economic and political power structures which exist both in and outside the community. While it can take a long time to fully identify and understand power structures within a community, a preliminary understanding can be obtained by asking key informants and community members a series of questions such as:

- Who are the most respected people in the community? Differentiate between women and men.
- Why are they respected?
- Who are regarded as the real leaders in the community?
- Are they well respected? Why?
- What role does politics play in community life?
- Who are the wealthiest people in the community?
- Where do they obtain their wealth?
- Who are the poorest people in the community? Why?

Additional questions can be added to determine the who, what, why, how and where of economic and political power in the community and the role this power can play in co-management.

- *Productive assets characteristics:* the ownership of fishing boats and gear, engine, level of investment, share system, sharing of catch.
- *Gender issues:* gender issues refer to the different roles, rights and responsibilities of men and women that are determined by social and cultural norms, and by biology. Gender issues include involvement in income-generation, control over the benefits of work, role in household work, time use, asset ownership, access to resources, rights in the household, rights in the community, and security and vulnerability of women.
- *Stakeholder perceptions:* stakeholder perceptions are how stakeholders think about resource conditions, threats to the resource, resource management, community and resource use conflicts, collective action, community issues, relations among people in the community, culture and beliefs.
- *Indigenous knowledge:* the knowledge held by people that is not scientifically based but comes from stakeholder observations, experiences, beliefs or perceptions of cause and effect and include folk taxonomy of resources, local knowledge of resources, and variations in knowledge.
- *Community services and facilities:* community services are services provided by individuals or organizations to support livelihoods of the community as a whole. Community facilities are the infrastructure that supports and facilitates the provision of those services, such as medical services, educational and religious facilities, public utilities, communication facilities, markets, transportation and other facilities such as hotels, restaurants and commercial businesses.
- *Market attributes for extractive uses of resources:* the characteristics of buying and selling coastal and marine resources that have been removed from the sea. Extractive uses refer to activities that take a resource without replacing it. Parameters include supply of the product, demand for the product, market prices, market structure, market orientation, market function, market rules, and market infrastructure and operation.
- *Market attributes for non-extractive uses of resources:* activities in which the resource is used, but nothing is taken or consumed as a result of the activity such as tourism and aquaculture. Parameters include demand for tourism activities, vulnerability of tourism market, characteristics of tourism stakeholders, supply of aquaculture, characteristics of aquaculture stakeholders, and aquaculture market structure.
- *Non-market and non-use values:* non-market value is the value of resource-related activities that are not traded in the market. Non-use values are the values that are not associated with any uses and include option value, bequest value and existence value.

A wide variety of field data collection and visualization methods is used for SEA (Box 7.17). Some of the methods are participatory and others are not. Some provide qualitative data and others quantitative data. The field data collection, emphasizing verbal modes of communication, involves a probing role for the researcher, a reactive respondent, the extraction of information, and sequential information flow. Field data collection methods include:

Box 7.17. Guiding Principles for Field Data Collection.

- Respect the stakeholders and communities;
- Clarify the objectives of data collection;
- Develop an interactive approach and communication between the team and the stakeholders;
- Recognize the limitations of information;
- Recognize informants' biases;
- Recognize and minimize biases of the team members including gender, education/discipline background, language, outsider priorities;
- Take detailed notes;
- Cross-check data;
- Create opportunities to reflect on learning;
- Recognize when to stop.

Source: Bunce *et al.* (2000).

- Observation;
- Semi-structured interviews;
- Focus groups;
- Surveys;
- Oral histories.

Many of these methods may occur simultaneously or sequentially.

Visualization and diagrams are pictorial or symbolic representations of information, and are a central element of participatory research. They allow non-literate and literate people to participate in the process as equals, facilitate the exploration of complex relationships, and generate collective knowledge. The visual mode of communication involves a facilitation role for the researcher, a creative and analytical role for the local stakeholder, the generation of local analysis and the cumulative flow of information. Techniques for visualizing and diagramming relationships in data include:

- Maps;
- Transects;
- Timelines;
- Seasonal calendars;
- Daily activity;
- Historical transects;
- Decision trees;
- Trend line;
- Process diagrams;
- Venn diagrams;
- Flow charts;
- Ranking;
- Classification.

These techniques are used to gather and present large amounts of complex information in a clear and concise, graphic and easily understood format. They also encourage interaction between the team and the informants; however, they rarely produce data that can be statistically analysed.

7.4.7. Legal and institutional assessment

The legal and institutional assessment (LIA) seeks to identify and analyse the organizations and governance structure for resource management in the community (Box 7.18). The LIA identifies the various resource users, stakeholders and organizations involved in resource management, analyses their roles in management, and evaluates the existing level of involvement of stakeholders in managing resources. The LIA identifies and examines the existing legislation, policies, regulations and programmes for resource management (fisheries, coastal management, marine protected areas, coastal ecosystems) at different levels of government (village, municipal, district, province, regional, national, international) and community (customary, traditional). The LIA identifies existing property right and tenure arrangements (formal and informal) in order to determine rights to access and use the resource, whether these rights are transferable, and the identification of the rules that must be followed. The LIA also identifies the existing political and economic power structures in the community to determine existing structures and what likely effects proposed changes in participation and governance will have. The LIA is crucial to the development of the management plan.

Organizations are groups of individuals bound by common purpose to achieve objectives. These include formal and informal decision-making and representative bodies, cooperatives and associations. Of concern are the organizations that formulate, supervise, monitor and enforce the various rights, rules and regulations governing coastal and aquatic resources. Institutions and agencies are those government bodies with responsibility for managing fish and coastal resources. These include Ministries or Departments of Environment and fisheries agencies.

Resource governance is the way in which resource users are managed by sets of rights, rules, social norms and shared strategies and includes enforcement mechanisms, such as policing measures and punishments. Resource governance can include:

- Formal and informal forms of resource ownership;
- Use rights and the laws that support these rights;
- The rules, rights and regulations that dictate how resources can and cannot be used.

Resource governance can be defined by formal organizations and law, by traditional or customary bodies, and/or by accepted practice.

Both community and external to the community institutional and organizational arrangements are identified and examined since, for example, national or international laws and policies can affect community level

Box 7.18. Institutional Analysis.

Institutional analysis is a participatory method which is used to identify existing legislation, policies and regulations for fisheries and coastal resource management at different levels of government and both formal and non-formal. It is used to identify existing property rights and tenure arrangements in order to determine who defines rights to exploit the resource, who has access to the resource, and the rules that must be followed. It is also used to evaluate the existing level of participation of resource users in managing the resource.

The approach to conducting an institutional analysis involves:

1. Collect secondary data on:

- Stakeholders;
- Organizations at the community level (mandate, functions, membership, structure, resources);
- Institutional arrangements at the community level (property rights/tenure, rules, regulations, boundaries, decision-making mechanism, monitoring and enforcement);
- Organizations/agencies above the community level (provincial/state, national, NGOs) (mandate, functions, structure, resources); and
- Institutional arrangements above the community level (provincial/state, national laws) (policy, legislation, regulation, programmes).

2. Complement and validate the secondary data collection by collecting primary data. A variety of participatory techniques and methods can be used. These include structured and semi-structured interviews, focus groups, resource mapping, historical timelines, flow patterns, case studies and Venn diagrams.
3. Collect and sort the data, focus on relationships between and among the various institutional arrangements and organizations for management.
4. Identify complementarities, conflicts, overlaps and gaps in the institutional arrangements and organizations which support or hinder effective management at various levels of government and within the community.
5. Identify what is needed to support management, such as new regulations, laws, organizations and enforcement mechanisms.
6. Recommend strategies for implementing patterns of relationships in space, time, flow and decision-making using various tools such as transects, maps, timelines, Venn diagrams and matrices.
7. Analyse the rules at operational, management and legislative levels.
8. Validate findings with the community to ensure accuracy and to fill in any data gaps.

A final report is produced containing descriptions, maps and figures that analyse the formal and informal fisheries and coastal resource management systems that operate in and around the community. The output can be used by fishers and government for dialogue and debate about resource management.

Source: IIRR (1998, pp. 118–130).

management plans. National administrative and economic development laws and policies are also examined since they may impact upon resource management and community development efforts.

There is a wide range of parameters which can be included in a LIA. These include:

- Political context: the political structure of the nation; the extent to and way in which stakeholders are represented; democratic processes and levels of representation.
- External to the community institutional and organizational arrangements (international, national, regional, provincial, municipal, village): government administrative agencies (mandate, functions, structure, resources); policies, legislation, regulations and programmes for resource management and environment, government administration, agriculture, and economic and community development; resource management strategies and programmes; non-governmental organizations (mandate, functions, structure, funding); surveillance, monitoring and enforcement; nested relationships between organizations and spheres of influence (complementarities, conflicts, overlaps, gaps which support or hinder effective management).
- Community institutional and organizational arrangements: identification of stakeholders; community organizations (mandate, functions, membership, structure, period of existence, resources, funding); boundaries (political, physical/natural, gear, customary, fishing area); property and tenure rights; rules and regulations (formal/informal, operational, collective choice, constitutional); decision-making and conflict management mechanisms; surveillance, monitoring and enforcement; compliance levels; nested relationships between organizations and rights (complementarities, conflicts, overlaps, gaps which support or hinder effective management).
- Incentives for collective action and cooperation among resource users.
- Extent of stakeholder participation.
- Extent of community-based management and co-management arrangements.
- Macroeconomic/political/sociocultural exogenous factors (natural calamities, political stability, peace and order, technological innovation, inflation, economic development, international agreements).

The level of detail of a LIA can range from a simple description of the existing coastal resource management system to a very detailed legal, economic and political analysis of the management system in terms of its impact on equity, efficiency and sustainability. Secondary data on organizations and resource governance can be obtained from official publications, including court records, official statutes and government reports.

In general the main methods of collecting primary data are semi-structured interviews and focus groups with key informants, such as government officials, organization officers, and other knowledgeable individuals involved in the organizations and governance. Some useful visualization techniques include:

- *Timelines* – to understand the history of organizations;
- *Organizational charts* – to represent aspects of the structure of the political hierarchy and the structure of organizations, as well as links between organizations and agencies;
- *Maps* – to illustrate areas covered by specific use rights;
- *Venn diagrams* – to illustrate organizational relationships.

Observations, surveys and oral histories can also be useful, particularly for assessing levels of stakeholder participation, surveillance, enforcement and compliance.

7.4.8. Problems, needs and opportunities assessment

The CO should have conducted a preliminary problems and needs identification early in the community entry and integration phase. A second problems, needs and opportunities assessment (PNOA) is conducted to update the information obtained earlier. The emphasis here is on identifying root causes of problems and agreeing upon them before solutions are identified and actions are taken. Since funds and resources to address problems will probably be limited, it is important to focus efforts.

As discussed earlier, workshops, meetings and discussions are held to help community members to identify and prioritize problems, needs and opportunities. A variety of methods can be used, including focus group discussions, problem trees and solution trees, problem analysis and ranking, brainstorming, and preference ranking (Box 7.19).

Following data collection under the REA, SEA, LIA and PNOA, data analysis is conducted.

7.4.9. Field analysis

It is important to conduct data analysis in the field as data are collected. The advantages of this include:

- The focus of the assessment can be adjusted in response to learning acquired in the field, making it an adaptive process;

Box 7.19. Preference Ranking.

Preference ranking allows the community to prioritize issues or options based on established criteria and individual preferences. This systematic ranking can be used to help the community identify the top concerns that should be addressed by their plan, ranking objectives of the plan, or selecting among interventions or activities within the plan. The approach involves:

- Holding a workshop of relevant stakeholders;
- Identifying and clarifying the issues or options and listing the options on a board;
- Establishing with the community the criteria for ranking or selecting among options;
- Asking each participant to score the options using a numeric system;
- Tabulating responses of the group members and summing the scores for each option;
- Developing a consensus among the group for selected options.

Source: IIRR (1998).

- The team's understanding of local conditions can be better used as not all their impressions and observations will have been recorded in a form that is easily reported;
- Stakeholders can participate in analysis, increasing their sense of ownership;
- Mistaken assumptions that may have influenced the design of the assessment can be corrected;
- The process of final analysis and reporting can be speeded up and facilitated so that the findings of the assessment can be quickly incorporated into the plans.

All team members (experts, community members, programme staff) should be involved and a workshop format is recommended to facilitate interaction among team members. Regular workshops should be held both for a specific component team and for all team members. The field analysis should identify key learnings about a parameter, cross-cutting issue, a particular problem or a specific question. Key learnings refer to issues identified or lessons learnt by the team that are essential to the objectives of the assessment or are needed to understand the context of the stakeholders and resource. Key learnings will be identified by team members informally discussing what they have learned. By comparing similar patterns and trends, new insights relevant to the goals and objectives can be identified.

Primary steps in the field analysis workshops include:

- Reviewing notes and questionnaires;
- Analysing quantitative data;
- Analysing qualitative data;
- Assessing the status of data collection and data and revising future data collection plans (Bunce *et al.*, 2000).

7.4.10. Final data analysis

Much of the data analysis, particularly of qualitative information, should have been completed during the field analysis workshops (Box 7.20). Therefore, the final analysis involves:

- Refining key learnings;
- Collecting and ordering data to illustrate key learnings;

Box 7.20. Basic Principles for Analysis.

- Involve all team members in the analysis;
- Prioritize quality, not quantity;
- Prioritize learning, rather than information;
- Do not modify the results to the end-users' expectations.

Source: Bunce *et al.* (2000).

- Presenting the key learnings in an accessible form to end-users;
- Validating the key learnings with stakeholders;
- Incorporating the key learnings in a useful report.

There are several critical steps involved in conducting final field data analysis:

- Compile all the information throughout the research.
- Prepare the quantitative data using statistical methods (simple calculations such as sums and percentages) and present in a graphic form.
- Assemble all team members for a final workshop to review, analyse and report the findings.
- Outline the final report and the type of report that is required depending upon target audience (for example, an executive summary for government leaders, a detailed description for scientists and agencies).
- Finalize key learnings by the team and match with assessment objectives to determine how they contribute. The team should synthesize results, share conclusions and discuss insights and recommendations.
- The information that has generated key learnings should be clearly identified by the team. The team should identify the various parameters examined and review how the information they collected contributes to the key learning.

7.4.11. Validation

Once key learnings, parameters and illustrations have been decided, it is time to validate these findings by presenting them to the stakeholders for comment. The findings should be in a clear and concise form, wherever possible using the visualization diagrams that the stakeholders have developed during the research. Long, verbal explanations or complicated tables of data generated during the research may be difficult for some people to understand.

Validation can take place in various forms:

- Small discussion groups with key stakeholders;
- Presentations to specific groups of stakeholders or interest groups;
- Presentations to groups of selected representatives of different stakeholder groups;
- Community meetings involving a wider range of stakeholders.

Discussions at these validation meetings should be recorded and the results incorporated into the final output. Although achieving a consensus is ideal, the more important thing is for the community members to be more aware of their problems and opportunities as they relate to the management of their resources. Consensus may not be desirable or achievable at the whole community level. It may be possible to gain consensus among community sub-groups. It will be important to document these differences. When stakeholders disagree with some of the results, the team must use its judgement to decide whether or not to change their results. Alternatively, it may be necessary to collect additional field data to clarify these discrepancies.

7.4.12. Final report

After the validation workshop, and if any additional information is collected as needed, the team should prepare the final community profile report (Box 7.21). The report should be circulated to the end-users and also presented again to different audiences as needed to provide information. The final community profile may be prepared in several different forms for different audiences, such as a policy brief for government officials and an illustrated form for illiterate community members.

Box 7.21. Suggested Outline for a Community Profile.

Executive Summary

Table of Contents
List of Tables
List of Figures
List of Acronyms
List of Appendices

Introduction
    Goal
    Objectives
    Scope
    Methods
    Sampling
    Short description of area
    Historical background
    Key learnings

Resource and Ecological Assessment
    Physical features
    Natural resources and habitats
    Technical aspects of production

Socio-economic assessment
    Resource use patterns
    Stakeholder characteristics
    Economic activities
    Gender issues
    Stakeholder perceptions
    Traditional knowledge
    Community services and facilities
    Market attributes for extractive uses
    Market attributes for non-extractive uses
    Non-market and non-use values

Legal and Institutional Assessment
    Political context
    External to the community institutional and organizational arrangements
    Community institutional and organizational arrangements
    Incentives for collective action and cooperation among resource users
    Extent of stakeholder participation
    Extent of community-based management and co-management arrangements

Problems, Needs and Opportunities Assessment

Management Issues and Opportunities

Recommendations

Acknowledgements

References

Appendices

8 Environmental Education, Capacity Development and Social Communication

R. Pomeroy.

Environmental education, capacity development and social communication (ECB) are integral parts of community organizing. The environmental education, capacity development and social communication activities can come before or at the same time as the other community organizing activities. However, due to their importance, they will be discussed separately from the other community organizing activities.

8.1. Environmental Education, Capacity Development and Social Communication

The issues of community-based co-management are generally complex and there is a need to promote environmental awareness in the community and to develop people's capacity to actively participate in the co-management programme. This includes the capacity of community members, as well as government officials and staff. (Throughout this section it should be kept in mind that the education and training of government officials and staff is as, if not more, important than that of community members.) The purpose of environmental education, capacity development and social communication is to empower people with knowledge and skills in order that they can actively participate in the community-based co-management programme, begin to take greater control over resource and economic and social problems and needs, negotiate a fair agreement, and increase their awareness and understanding of fisheries resources and their management. Through ECB, community members and government officials and staff are able to better understand the need for co-management, the approaches to co-management, and their individual and collective roles in co-management (Table 8.1). In some cases, the community and government may need to be convinced of the need to protect and manage their own resources and for co-management. While ECB is a continuing activity throughout the co-management programme, it should be noted that it is important to start the ECB activities as soon as possible in order to empower people with knowledge and skills so that they can actively participate in the co-management programme.

Table 8.1. Environmental education, capacity development and social communication.

Stakeholder	Role
Fisher/fisher organization	• Participation • Input as to education and capacity development needs
Other stakeholders	• Participation • Input as to education and capacity development needs
Government	• Support • Participate in activities • Participate in education and training
External agent/CO	• Organize and conduct training and capacity development activities

Activities aimed at increasing awareness, knowledge, skills and institutional capacity, such as environmental education, capacity development and social communication, are sometimes taken together under the term 'social preparation'. Social preparation has several functions in co-management:

• Reducing social conflict and resource impacts;
• Creating positive change in values and behaviour towards the environment;
• Gaining support for co-management;
• Increasing knowledge and skills of fishers and other stakeholders;
• Fostering participation in community-based co-management;

- Enabling community members to assert their rights to use and manage its resources.

The ultimate goal of social preparation is to achieve behaviour and attitude changes so that resource use and management and the co-management programme can be sustainable. Social preparation is focused on building a constituency for co-management through a critical mass of people in the community who are environmentally literate, imbued with environmental ethics, shared responsibilities, and shared actions towards the sustainable management of aquatic resources (DENR *et al.*, 2001b).

It should be noted that social preparation activities alone will not cause people to change unsustainable practices and behaviour. There need to be several actions operating concurrently, such as changed community values, availability of alternative behaviours, and possible sanctions for unsustainable activities.

8.2. ECB Activities

Environmental education, capacity development and social communication are individually distinct but complementary activities.

Environmental education introduces environmental concepts and principles related to coastal and aquatic resource issues, and empowers the community with information and knowledge in order to take the appropriate action to address the issues. The success of aquatic resource management depends on the level of the community's awareness and knowledge of their coastal and aquatic environment. Environmental education activities are directed towards the development and enhancement of resource management capabilities of individuals and organizations through formal and non-formal education and skills development training (Juinio-Menez *et al.*, 2000). Environmental education can build consensus, clarify perspectives and interests about issues, generate a receptive context for change, get people to help carry out activities, help monitor change and create a long-term commitment in the community.

Capacity development provides skills and institutional capacity for fishers, resource user organizations, local-level government officials and staff, and other stakeholders to take an active role in co-management. Capacity building often implies that activities are carefully planned and executed, and that they follow a clear plan. In reality, capacity building often involves more experimentation and learning. For this reason, the term capacity development, which implies an organic process of growth and development, is more appropriate than capacity building (Horton, 2002). Capacity development can be defined as

> the process by which individuals, groups, organizations, institutions and societies increase their abilities to: (1) perform core functions, solve problems, define and achieve desired objectives over time; and (2) understand and deal with their development needs in a broad context and in a sustainable manner.
>
> (UNDP, 1998)

This definition highlights two important points: (i) that capacity development is largely an internal process of growth and development, and (ii) that capacity development efforts should be results-oriented.

Local capacity is built in order to:

- Make local resource users, groups and organizations, fishing communities and the local government unit charged with fisheries management more capable of performing this task;
- Make local resource users, their organization leaders, local government officials and staff, and other stakeholders able to undertake their roles and responsibilities in co-management;
- Improve the quality of fisheries management taking place at the community level.

Social communication generates an on-going flow of information and dialogue between the CO and the community members, and among the community members themselves in order to have informed decision-making and to face change. Social communication initiatives can promote social discussions about problems, opportunities and alternative courses of action, including co-management, for the community. Social communication initiatives are very different from education initiatives. They do not merely aim at 'passing on a message about an issue' but at promoting its critical understanding and appropriation in society (Borrini-Feyerabend *et al.*, 2000).

ECB activities should involve as many of the sectors of the community, including government, as possible in order to build up a critical mass of local people with a common understanding of co-management and aquatic resource management. Efforts should be focused on cultivating potential local resource persons who could effectively conduct ECB activities on their own (e.g. local teachers to their students and other teachers) and in the process disseminate information to even more members of the community, leading to the greatest positive impact in the shortest period of time (Juinio-Menez *et al.*, 2000). It is important to monitor and evaluate the effectiveness of ECB activities, including changes in the community's attitude to the need for co-management.

8.3. Environmental Education

Environmental education is a critical ingredient in the transformation of community members into active partners in co-management. Environmental education empowers people and improves their environmental awareness through knowledge (Boxes 8.1 and 8.2). The effectiveness of environmental education activities is manifested through:

- Concrete action taken by local co-management advocates towards various resource management issues and dissemination of information to a greater number of community members;
- Direct participation of local communities in concrete co-management

Box 8.1. Community Education on San Salvador Island, Philippines.

An ongoing education and training programme used formal and informal approaches and various resource persons. Monthly education programmes, which used slide shows, role-playing and lectures, relied on basic ecological and environmental concepts and on feedback from survey results. Children's outings and an environmental drawing contest were also effective.

A field trip to another community sanctuary project on Apo Island, in the province of Negros, by seven San Salvador residents was a major educational activity. It was supported by continual informal contact between the two field-workers and the community. As a result of the trip, San Salvador fishers were inspired to form the Lupong Tagapangasiwa ng Kapaligiran (LTK) or the Environment Management Committee. The LTK actively educated and motivated other residents to participate in the overall community-based management programme.

Source: Buhat (1994).

Box 8.2. Community Environmental Education in Bolinao, Northern Philippines.

The Community-based Coastal Resource Management programme in Bolinao, Pangasinan, Philippines, was implemented by the Marine Science Institute and College of Social Work and Community Development of the University of the Philippines and the Haribon Foundation for the Conservation of Natural Resources. The overall vision of the programme was to facilitate the sustainable development of the coastal zone with the local communities acting as coastal resource managers. The foundation of the programme was through community organizing and environmental education. The programme acted to empower the communities through knowledge and skills development and advocating for active community participation in coastal resource management.

Heightening the environmental awareness of the people of Bolinao was one of the long-term goals of the programme. However, the environmental education activities undertaken were geared primarily towards the development and enhancement of the resource management capabilities of people's organizations through formal and non-formal education and skills development training. The environmental education activities were integral to the community organizing process.

Programme activities were focused on the people of four coastal villages, as well as local science educators and representatives of the local and regional governments regarding the coastal environment and its effective management. This was done in order to facilitate the building up of a critical mass of local people, all having a common understanding of coastal resource management. Through the environmental education activities, conscious efforts were made to cultivate potential local resource persons who could effectively conduct environmental education activities on their own (e.g. village leaders to other adjoining villages; science teachers to their students and other teachers); in the process disseminating information to even more members of the local community.

Environmental education extended beyond providing and gathering information on various aspects of coastal resources management. Consistent with the participatory nature of the programme, opportunities for involvement in actual resource management activities (e.g. coastal mariculture, mangrove reforestation, coastal planning) were integral parts of the process of instilling values and skills for coastal resource management. Developing the ability of local communities to make informed decisions through careful evaluation of biophysical, socio-economic and practical considerations attendant to a range of resources management initiatives was another major objective of the environmental education programme.

Source: Juinio-Menez *et al.* (2000).

initiatives such as planning, implementation and monitoring (Juinio-Menez *et al.*, 2000).

8.3.1. The environmental education plan

An environmental education programme should be based on a plan that ensures that environmental education activities have a connection to the community and its goals for sustainability, emphasize the link between local resource-use activities and the quality of the environment, and ensure that it is relevant to people's lives. Too often, environmental education is not targeted at groups in the community who are most important in resource use and management and in participation. The plan should consider the issues and needs of the whole community (Boxes 8.3 and 8.4). Environmental education activities should be developed collaboratively among the community, external agent and government. The educational needs from the community's perspective may be different from the perspective of the external agent.

Training needs analysis should be conducted through questionnaires or by interviewing community members to determine the appropriate topics, inputs and methods for the training. Decisions about the selection of topics, planning and implementation of training activities should be made with all partners. The environmental education programme should be based on accurate local information. The Resource and Environmental Assessment, Socio-economic Assessment, Legal and Institutional Assessment and Problems, Needs and Opportunities Assessment all provide this type of information for the formulation of an environmental education plan (see Chapter 7).

For the environmental education programme (or capacity building and social communication), it is important to develop criteria for success, as well as mechanisms for monitoring and evaluating the effectiveness of environmental education programmes, including changes in the community's attitude to the need for resource conservation and co-management. Often resistance or outright opposition to participation is a useful indicator that the co-management arrangements being developed are not going to meet the perceived needs of the people in their present form. Instead of being taken as failure, this can be a cause for the effort to find out the reasons for the resistance to participation. This should then lead to a review of the

---

Box 8.3. Steps in Designing and Implementing a Public Education Programme.

1. Analyse the local context and define the major coastal issues

- What is the scale and significance of the problems?
- Are there important social, economic or ecological dimensions to each of the problems?
- Have technical causes been identified?
- Have technical solutions been identified?

2. Identify target audiences

- Who has a direct stake in co-management?
- Who will be directly affected by co-management?
- Who uses coastal resources?
- Who decides how coastal resources will be allocated?
- Do these audiences have special information needs?
- Do they have a unique perspective or knowledge of coastal issues?

3. Identify the message and programme content

- What is the education programme attempting to accomplish?
- Are the target audiences directly affected by resource deterioration? In what ways?
- What role will these audiences play in implementing possible solutions?
- What do people need to know or feel strongly about in order to act?

4. Select techniques and media

- How do the various target audiences keep informed?
- How accessible are the target audiences? Are there convenient distribution networks?
- Is the educational message simple? Or complex?
- How much money is available? What are the local resources (both financial and human) that can be drawn upon?

5. Evaluate the programme

- Did the information reach the target audiences?
- Was the message accurately conveyed by mass media?
- Did people understand the information?
- Was there a response from the target audiences?

Source: Lemay and Hale (1989).

---

environmental education activities and possibly also a re-examination and redesign of programme objectives (Claridge, 1998).

Conversely, it is always a good idea to seek clarification of why communities, or groups within communities, are willing to participate. Their motives may not, for example, always be consistent with the sustainable resource management objective of the programme. Even in the usual situation of motivation through a desire for sustainable development, understanding the motivation of participants can provide a guide to ways in which involvement can be increased. Again, the environmental education programme activities are closely linked to community attitudes and any perceived problems with

Box 8.4. ECB in the Marine Reserve in Trao Reef in Khanh Hoa Province, Vietnam.

Trao Reef Locally Managed Marine Reserve is a 3-year project implemented in Van Hung Commune, Van Ninh district, Khanh Hoa province, both funded and facilitated by the International Marinelife Alliance (IMA) Vietnam. The project supported the management of the Xuan Tu community marine reserve and assisted in restoration activities. Van Hung commune, particularly Xuan Tu village, is the main area where this project is being implemented. Trao Reef is composed of fringing reefs that run along or close to the coastal shoreline. It ranges from a depth of 2–3 m near shore to 6–7 m further away. Part of Trao Reef is even exposed at low tide. It has a remaining coral coverage of 40–60%. Recent marine field surveys have shown that, since 1980, there has been an average 10% decrease in the availability of marine resources in the area. Some species often found and caught in the past, such as abalone, sea cucumber and grouper, are now rare.

The key approach of the project is community-based participation. IMA Vietnam facilitated the establishment of a community-managed reserve. It is protected by the core group, that is elected by and representing the community and working following community-established regulations. Increasing the level of awareness and understanding of communities on the benefits of resource management is a primary concern of the project. To achieve this, the project had coordinated with Vietnam National Television, collecting information and facilitating their film making on coastal resources management activities in the project site. School competitions to describe the environmental problems and propose solutions were held. Subsequently the Le Hong Phong secondary school worked with IMA Vietnam to develop an environmental education programme for incorporation into the current school curriculum. IMA will provide videotapes and other relevant materials and support other activities for raising awareness among the pupils.

In the villages, activities such as the international coastal clean-up days were held attracting thousands of people. Collecting and organizing trash has become a good habit of the local villagers. Several clean-up-oriented activities have promoted awareness-raising about environmental issues such as trash. Over 600 people participated in these activities led by the Veterans' Association, Women's and Youth Union and Advocates. Children in primary school and women are primary participants. They collected plastics, tins and other non-biodegradables as well as accumulated dead vegetation. Some larger trash items were deposited in the Commune's landfill. Other trash was burned on site. One of the significant results gained from the above activities is the formulation of the local village trash collection group. It has been put into good operation with obvious regulation and the contribution made from the community members.

To complement the clean-up activities, various competitions on writing and singing were also held in the village. IMA Vietnam also facilitated study tours to Trao Reef and set up dialogues on marine biodiversity and resources conservation in relation to lobster aquaculture. Youth, women, fishermen and teachers of the local primary school participated in these dialogues. With these activities, the people had a chance to see the coral reefs and the impact of the protection efforts. It is obvious that these study tours and dialogues raised the awareness of the villagers and guided their action in marine resource protection.

All of these activities facilitated an atmosphere of cooperation, collaboration and success. A major learning is that co-management of resources is an evolving process where sustained training, facilitation and support is needed. Environmental education work provides an opportunity for sharing learnings that eventually informed local decision-making.

Source: International Marinelife Alliance Vietnam (nd).

motivation should feed back into the contents and methodology of the environmental education programme (Claridge, 1998).

8.3.2. Entry level approaches

Environmental education activities start in the community entry and integration phase of the co-management programme. The target groups at this early stage include potential leaders, key informants and government staff in the community who are showing an active interest in the process and who could disseminate information to others. The activities are non-formal and may include one-on-one and small group discussions about environmental issues in the community (Boxes 8.5, 8.6 and 8.7). The environmental education

---

Box 8.5. Small Group Discussions.

Small group discussions should be evocative. Participants should be given time and opportunities to voice out their personal views. It is important that the exchange of ideas be as spontaneous as possible, especially when getting information regarding the level of awareness of the people and when gathering the insights of the local people on their role as co-managers of the resources. Refrain from using recording instruments such as tape recorders as these usually make participants hesitant and uneasy.

Source: Juinio-Menez *et al.* (2000).

Box 8.6. Introduction and Expectation Check.

Icebreakers
An easy way to help a facilitator get training started is to facilitate an exercise that allows participants to have fun while getting to know each other. One example is an activity called 'A Picture of Myself'. To do this, the facilitator hands out pieces of paper and thick markers to all the participants. Participants are asked to draw a picture of themselves, however they like. Then, at the bottom of the drawing, the participant writes three pieces of information about themselves that they will share with the group, e.g. where they come from, what they like. When everyone is finished, each participant shows the group what they have drawn and explains this.

Choosing an introduction activity for participants is important for any training, especially if they do not know each other. The facilitator should allot enough time for introduction and icebreakers so that the participants will feel comfortable at the start of the training.

As a follow-up to an introduction, the facilitator needs to check the expectations of the participants. To do this an activity called the 'Sea of Expectations' can be introduced. For example, the facilitator provides participants with paper cut outs of fish, seagrass and dynamite and markers. Participants are then told that the fish cut-outs represent one's expectations about the training, the seagrass symbolizes one's background/experiences, and dynamite represents fears or worries related to the training. Participants are given 10 minutes to write down their ideas, at which point the 'sea of expectations' is pinned on to the whiteboard. Participants each take turns sharing their responses to the group. The activity ends with the facilitator's summary of responses.

Source: Rivera-Guieb and Marschke (2002).

Box 8.7. Sharing.

Values are changed by talking, working and sharing together. Change can happen to community members and to the community organizer as they immerse themselves in the project. Learning and sharing take place through informal and spontaneous sharing. This kind of sharing does not take place in a formal setting but in any place, any time in the community. It can take place while people are on board a passenger bus or a boat, celebrating a fiesta, or having a talk in the market. People can talk about their feelings, their impressions of the day and the progress of the project in an informal way that resembles a community-wide assessment. There is a continuous sharing of opinions and ideas and this is crucial to community organizing work.

Source: Arciaga *et al.* (2002).

---

activities involve deepening awareness of the target groups about environmental issues and possible solutions (Juinio-Menez *et al.*, 2000).

In addition to community members, government staff at local levels will possibly need environmental education (as well as capacity building and social communication). This can sometimes cause difficulties as government staff may not be willing to admit their lack of knowledge or to put effort into learning about community groups they may feel to be ignorant or uneducated. It may be necessary, in some cases, to have separate training for government staff or to use alternatives to conventional 'classroom'-type training. An alternative may be an exchange visit to an area with successful co-management or on-the-job training. It may be useful to involve community members in training activities in order to show their knowledge, especially if field-type training is used (Claridge, 1998).

8.3.3. Training

The primary method of environmental education is training. Training may involve formal (training sessions, workshops, lectures) and non-formal (small groups, exchange visits, peer-to-peer discussion, plays, one-on-one contact) educational methods. Training needs to be a continuous process throughout the programme.

A variety of environmental education training modules can be prepared (Box 8.8). Juinio-Menez *et al.* (2000) have identified seven training modules:

1. *Marine Environmental Situationer.* This module provides an overview of the current situation of the marine environment and its related resources. Environmental, socio-economic and sociopolitical factors are discussed to provide the participants with a holistic view of the issues and to enable them to realize the complexity of marine environment management. The common causes of environmental problems are also briefly discussed. The preparation of a local situationer is intended to heighten the awareness of the participants

---

Box 8.8. Environmental Education in Belize.

The Friends of Nature (FON), an NGO based in Placencia, co-manages the Laughing Bird Caye National Park and the Gladden Spit and Silk Caye Marine Reserve, with government. As part of its staff, FON has a full-time environmental educator. In its strategic plan, FON has identified education and outreach as one of its primary activities. FON has specifically identified the following interventions as part of this activity:

- Student environmental education materials;
- Teacher environmental education lecture materials;
- Environmental education lectures at local schools;
- FON information materials;
- Resource user (fishers, guides) environmental education materials;
- Tourist environmental education materials;
- Lecture series on environmental matters for local community;
- Tourism professional orientation;
- Fishermen training courses;
- Development of environmental library at FON's office and in each school and community.

To date, FON has produced information brochures for visitors to the Park and Marine Reserve. The environmental educator has initiated a primary school environmental education curriculum working with the teachers. Training on biodiversity and coastal management has been provided to the dive and fishing tour guides from the area.

Source: Pomeroy and Goetze (2003).

---

on local marine environmental issues and motivate them to initiate environmental management actions in their locality.

2. *Basic Marine Ecology.* This module introduces several basic ecological concepts and processes. Understanding these concepts is a requisite for identifying and discussing the different ecosystems found in the country. The features and characteristics of the marine environment are likewise discussed to enhance the participant's appreciation of the complexity of this environment. Connectivities of the different coastal ecosystems are presented to contextualize the need for the integrated management of those ecosystems. Integration of local knowledge and perceptions regarding the natural resources are emphasized.

3. *Marine Biodiversity.* This module discusses the diversity of the marine environment. The components of biological diversity are discussed to give the participants an understanding of the importance of biodiversity. The exceptional diversity of individual coastal ecosystems is discussed to encourage the participants to initiate coastal resource management in their own areas.

4. *Sustainable Use of Natural Resources.* This module discusses the concept of carrying capacity and emphasizes that the earth's resources are finite and that Mother Nature has set natural limits to maintain ecological balance. Some causes of environmental problems are mentioned so that the participants may relate their personal activities to the observed environmental problems. The module explores the different possibilities by which the participants can determine how they can contribute to the sustainable management of their coastal zone.

5. *Resource Management Options.* This module provides an overview of the different options available for marine resources and ecosystems management. Fishery regulation, environmental intervention and community-based approaches are discussed to facilitate the identification by the participants of the different environmental issues facing them and of the possible courses of action that they can take. By the end of the session the participants should already have an idea about which local environmental issues they will prioritize and what management options to implement.

6. *Marine Protected Areas.* This module introduces the concept of the marine protected area (MPA) as a resource management option. The significance of MPA is discussed for the participants to fully understand its important role in coastal resources management. Sanctuaries and reserves, the two major components of an MPA, are also discussed. The biophysical, socio-economic and practical considerations that need to be considered during the planning and implementation of MPAs are likewise discussed. Different case studies are presented to give the participants an idea on the process of establishing and managing MPAs. Lessons are drawn from the experiences of different institutions engaged in MPAs. At the end of the module, the participants are expected to formulate a conceptual MPA framework and outline the attendant management plans.

7. *Coastal Development Planning.* This module introduces coastal development planning as a resource management tool. It highlights the importance of multi-sectoral participation in the formulation of an integrated plan. Being multi-sectoral, it addresses the environmental, socio-economic and sociopolitical issues existing in the area to ensure that sustainable and equitable development of the coastal zone is achieved.

Additional topics can include:

- Sustainable exploitation;
- Habitat improvement and enhancement;
- Ecological relations between land and sea;
- Relationship between people and natural resources (value and attitude development towards responsible stewardship);
- Environmental legislation;
- Enforcement and compliance;
- Conflict management;
- Gender roles in resource use and management.

8.3.4. Trainers' training

Environmental education efforts should be focused on cultivating potential local resource persons who could effectively conduct ECB activities on their own and in the process disseminate information to even more members of the community. This 'trainers' training' can target community leaders and teachers.

Environmental education training for community leaders targets selected local leaders who have the capability and respect in the community to become effective local resource persons on co-management and resource management. This is done to ensure the continuity and expansion of co-management in the area beyond the programme's duration. The training is designed to equip the participants with the basics of popular education methods and to generate individual and collective action towards the implementation of environmental education activities in the community (Juinio-Menez *et al.*, 2000).

Teachers' training workshop is a specialized training on the marine environment and on co-management for local science educators within a community. This training focuses on information rather than on the methods for conducting an environmental education training. The teachers can become catalysts in raising the environmental consciousness not only of the students but also of other teachers in the area. Major components of the training can include the state of the marine environment in the nation and locally, principles of marine ecology, fisheries resource management, integrated coastal management, and co-management. This training can enhance the knowledge and understanding of local educators regarding the marine environment and its conservation. It can also facilitate networking among fisher organizations and local educators. Laboratory exposure and field trips can enhance the lectures (Juinio-Menez *et al.*, 2000).

Monitoring the impact of the training is crucial to sustaining the momentum created by the workshops. A questionnaire can be administered in the community to evaluate the impact of the training workshops on the community.

8.3.5. Guidelines for training

Some guidelines for training sessions and workshops in communities include:

- Schedule the training at a time and place convenient to the participants.
- Facilitators should be assigned to run the training.
- Documenters should be assigned to record the training and important results and recommendations.
- Local and outside resource persons can be assigned to discuss/facilitate some topics during the training.
- Information should be presented in an accessible and easily understood format to broad audiences to raise awareness about the aquatic environment.
- Start from what people already know; dig for indigenous knowledge and add scientific knowledge when this can complement existing knowledge in a useful way.

- Facilitate sharing of indigenous knowledge between older and younger generations or between original community residents and more recent migrants.
- To explain new ideas and concepts, look for similar situations or concepts outside fisheries but within the everyday life experience of the community.
- Choose for each subject or topic the most appropriate method or form of training and the most suitable trainer.
- Technical terms should be explained and expounded on by using simple words.
- Use both formal (training sessions, workshops, lectures) and non-formal (small groups, exchange visits, peer-to-peer discussion, plays, one-on-one contact) educational methods.
- Non-formal methods are found to be best as they permit participation and interaction and encourage personal contact between the experts and the community, between peers (such as fisher-to-fisher), and among the stakeholders themselves (Box 8.9).

Box 8.9. Exposure/Study Trip.

A good example of an informal training is an exposure trip to a site that is somewhat more advanced (for instance in terms of participatory resource management) and that can serve as an example. The exposure trip provides first-hand knowledge regarding co-management and resource management. The exposure trip can also establish linkages with various NGOs and resource user organizations. Also the invitation of a fisher-man or -woman from such a site or organization to share with the group or community can be an effective way to raise confidence and bring in new ideas and inspiration. Communication and sharing among peers is often more effective than the transfer of knowledge or experience between persons with a different cultural, educational or professional background. Exposure can help to develop a vision or clearer direction (inspiration). The experiences of the people on the site can serve to guide the process (avoid mistakes, imitate successful processes); it can also help to realize that others have or had the same type of problems as experienced by the community members but they were able to overcome them. Feedback and reflection sessions should be conducted after the exposure trip to assess the participants' learnings and insights. These lessons are vital for sustaining the momentum created by the activity.

Source: Juinio-Menez *et al.* (2000).

- Evocative processes can be effective in encouraging active participation from the participants.
- The sharing of indigenous knowledge can play a key role in effective non-formal education programmes.
- Be creative. Games, acting and role playing can all be used in training and are fun (and, therefore, are remembered).

- Games can be used to explain resource management issues.
- Lectures alone should not be used. This is the form academically trained people are most familiar with and they will often automatically apply this method themselves. Be creative and try to think of other, more participatory ways to transfer knowledge or discuss a topic (Box 8.10).

Box 8.10. The Importance of Training and Capacity Building in South Africa.

Experience with co-management in South Africa highlights the importance of incorporating a capacity building component into the co-management process. There is a necessity for the resource users to obtain an understanding of the concepts and principles of sustainable resource use. Training and capacity building interventions included teaching basic life skills such as literacy, business and organizational management, including the operation of committees, and the principles of resource management. The most effective process of building capacity is through 'learning-by-doing' that involves resource users in research and monitoring activities. Arranging exchange visits between communities engaged in co-management or wishing to embark on co-management seemed to be particularly effective.

In several of the cases in South Africa, failure to allocate sufficient time and resources to developing institutional and human capacity was identified as one of the main obstacles to implementing effective co-management. Lack of effective community structures and skills training was considered to be a contributory factor to the failure of one project. Building requisite human and organizational capacity among communities and government departments at various levels was found to be an integral component of a co-management process in South Africa.

Source: Sowman *et al.* (2003).

- Use as many visuals as possible. Photos and videos add to learning.
- Realize that skill building requires practical exercise and repetition for a person to become skilful.
- Drawings, posters and slides can be very effective in meetings, training sessions and workshops for mostly illiterate people. Be aware that villagers can interpret pictures in a different way than was intended by the person who made the drawing. Posters that are going to be used repeatedly should be pre-tested for a small group of people from the target group before being reproduced and used on a larger scale.
- Whenever possible, all examples should depict local settings and situations to make the topic easier to appreciate.
- In general, we remember more of something when we have been more involved, or when more of our senses have been involved.
- We remember:
  - 10% of what we read;
  - 20% of what we hear;
  - 30% of what we see;

- ◦ 50% of what we hear and see;
- ◦ 70% of what we say;
- ◦ 90% of what we say and do.
- Small group meetings where there is a high level of participation by the participants can be supplemented by an occasional more formal presentation.
- Ensure that all materials needed for the training are available in advance.

8.3.6. Other methods for environmental education

In addition to training, a variety of other methods can be used for education (as well as capacity building and communication) (Green, 1997; Juinio-Menez *et al.*, 2000):

- *Participatory Research.* As discussed previously, an effective means of environmental education is to involve community members in activities of the REA, SEA and LIA to develop skills, utilize indigenous knowledge, and understand functions and values.
- *Videos.* Both outsourced and locally produced videos can be used. Outsourced videos on a variety of topics can be obtained from other external agents or projects in the country or internationally. Locally produced videos can be very effective since they address local issues with local people. Once a clear plan is developed and some rough editing tools are obtained for the computer, these videos are relatively easy to produce. Local people can often better relate by seeing themselves and their neighbours in the video. Videos can be adapted to a variety of audiences.
- *Television.* Usually expensive and sometimes not accessible by the community. It may be possible to obtain an outsourced video or produce a local video and request the local television station to show it. It may also be possible to interest the national and international television networks, such as CNN and BBC, in local stories which can provide good exposure to the community and the issues.
- *Radio.* A very effective method of communication to reach a large area if most community members have access to radios. Local radio stations can be requested to provide time for a regularly scheduled programme. The radio show can include interviews with local and outside experts, debates with stakeholders, documentaries, news, talk show where people phone in and information to stimulate discussion. Radio dramas have proven to be very effective.
- *Newspapers.* Local newspapers are a sure way of accessing local influential people. Newspapers may have or may be asked to develop a section on the environment. Journalists can be educated about environmental issues which will give them a greater interest in the subject. Good local stories can be provided to the national press.
- *Newsletters.* Newsletters can be prepared and distributed on a regular basis to inform about the programme activities, issues and upcoming events.

- *Fact Sheets/Flyers.* Short and simple, fact sheets and flyers can be used to provide information on a specific topic or issue.
- *Comics.* These can be used to disseminate information at the local level, especially to those who are illiterate. Comics must be short, be in the local language, and have a clear point relating to daily life with plenty of illustrations. Comics can be produced very cheaply and one copy can be shared by a whole family.
- *Posters/Calendars/Fixed Exhibits.* Very effective in the correct location. A calendar will stay on a wall all year. A colourful poster with a clear photo or illustration can deliver an important message. Fixed exhibits can be located in often-visited locations such as a town hall, school or market.
- *Picture Stories.* Picture stories can be presented in the form of flannel boards or flip chart drawings or some variation of these. They are illustrations of problems and solutions which can be put in sequence to tell a story and can be altered and added to in response to community feedback. Simple, colourful pictures can be very effective in helping people remember the key message of a presentation.
- *Information Kits.* A collection of materials on various topics organized and collected in a folder or binder. Larger information kits can be used for school education programmes and can include publications, handouts, presentation materials, posters, laboratory equipment and other materials.
- *T-shirts.* Sometimes useful in awareness raising and as a prize/gift for participation. They can often be a good conversation starter. Different t-shirts can be produced for different stakeholder groups and messages.
- *Sponsoring Events.* Local events can be sponsored to provide relevant environmental information to the general public. These can include such events as Earth Day celebrations, Coastal Clean-Up, sports competition or a village feast/celebration. Theme nights about the sea, such as an environmental concert where the fishers write and sing the songs can be effective. Other events may include plays, dramas and contests, or high profile activities such as beach clean-up or tree planting.
- *Street or Village Theatre.* Street or village theatre uses storytellers, theatre groups, clowns, dancers and puppets to inform people about an issue by telling a story. The presentations use imagery, music and humour to raise people's awareness of an issue that is affecting them. Local people can be encouraged to join in and play a part in the presentations. The presentations can be filmed or recorded for radio and thus made available to a wider audience.
- *Issue-based Information Campaigns.* Information campaigns can be launched in strategic locations to enhance the environmental awareness of the people. Depending upon the output of the information campaign, a forum can be held to strengthen collaboration among different support groups. Lectures, leaflets, photo exhibits, film/video showings and testimonials can all be utilized. If possible, networking with different support groups (e.g. international scientific communities, NGOs) should be facilitated to further strengthen the advocacy.
- *Community Consultations.* Community consultations are conducted in order

to disseminate information, gather and validate vital information and lobby support for the project. Visual aids and resource persons can be utilized.

- *School Curriculum Development.* Targeted curriculum development in primary and secondary schools on such topics as coastal ecology. This will often need to be coordinated with the Department/Ministry of Education and the local school district. Curriculum modules can be prepared which include lectures, readings, laboratory and field experiments, science projects, site visits, videos, PowerPoint presentations, computer exercises and other educational materials. Teachers should be involved in the development of the curriculum module. Scholarships can be provided to students through local competitions.
- *Recreational/Site Visits.* Local people, government officials, government staff and others can be invited on an outing for a picnic and day at sea. Participants can snorkel and explore the coastal ecosystems. These can be good fun and provide informal education.

8.4. Capacity Development

Complementary to environmental education, there is the need to develop the capacity for individuals and organizations to effectively participate in co-management. Capacity can be defined as 'the ability of individuals and organizations or organizational units to perform functions effectively, efficiently and sustainably' (UNDP, 1998). Capacity development includes understanding what co-management is and how to organize and participate in it, communicating with other stakeholders, dealing with administrative and business matters, and participating in negotiations. Capacity is a continuing process and is the power of an individual or organization to engage, in this case, in co-management.

The objective of capacity development is not to supply a product or service but to foster the development of specific individuals and organizations (Box 8.11). Capacity development is often needed to raise an organization's performance level, which is reflected in its efficiency (minimizes costs), effectiveness (achievement of its goals) and sustainability (relevance and acquiring resources for operation). The core capacities of an organization or community consist of:

- Defining and analysing the environment or overall system;
- Identifying needs and/or key issues;
- Formulating strategies to respond to or meet needs;
- Devising or implementing actions; assembling and using resources effectively and sustainably;
- Monitoring performance, ensuring feedback and adjusting courses of action to meet objectives;
- Acquiring new knowledge and skills to meet evolving challenges.

Most capacity development efforts focus on only one or a few of these critical capacities needed by an organization, on the assumption that the

Box 8.11. A Framework for Capacity Development.

The Caribbean Natural Resources Institute (CANARI) has developed a framework for capacity development that contains seven main elements that organizations should focus on, illustrating the breadth of capacity development beyond training:

- World view: vision and mission guiding capacity requirements;
- Culture: an organization's distinctive climate and way of operating;
- Structure: roles, functions, positions, supervision, reporting, etc.;
- Adaptive strategies: ways of responding to changing environments;
- Skills: knowledge, abilities and competencies for effective action;
- Material resources: technology, finance and equipment required;
- Linkages: relationships and networks for action and resource flows.

Source: Krishnarayan *et al.* (2002).

improvements brought about in these capacities will lead to improvements in the performance of the organization as a whole. This assumption is seldom tested, however. The capacity development efforts should be tested periodically in order to provide a basis for improving future capacity development efforts (Horton, 2002).

A key issue in capacity development is what is referred to as 'social capital'. It is important to recognize that the whole social community is more than the sum of its individual parts. People form relationships that fulfil a number of social needs such as communities of common interests, mutual obligation, care, concern, interest and access to information. These can be considered as networks of norms and trust which facilitate cooperation for mutual benefit. Social capital facilitates a process of learning through interaction. This social capital is critical to achieve collective action and to prosper and sustain a social, economic and institutional environment that is ready to adapt and change. The social networks can be horizontal (across the community) to give communities a sense of identify and common purpose or vertical (government to community to individuals) to broaden capacity and support.

8.4.1. Levels of capacity development

Capacity development efforts may focus on different levels. Capacity for co-management can take place at three levels:

- Individual;
- Organizational;
- System or enabling environment.

These three levels are nested within each other and there is regular interaction to form a whole. Capacity development efforts need to address challenges at various levels in the community, as well as externally.

The *system or enabling environment* level includes two distinct but complementary capacity development activities. First, this level includes capacity development for the broader co-management programme and associated activities at the community level. This includes developing capacity to prepare for negotiation, developing a common vision, negotiating plans and agreements, organizational representation, conflict management, and monitoring and evaluation. Second, this level also includes capacity development at the community level to support advocacy and networking for the enabling environment for co-management including policy, legal, regulatory, management and resource dimensions. The focus of this level is on the government or public sector, but may also include private companies. It has a national or regional scale and is multi-sectoral. It is to provide a political voice so that economically and socially disadvantaged user groups and communities can be considered in decisions and processes that relate directly to the resource and the well-being of the group or community.

Capacity development at the *organizational* level, such as a community-based fisher organization, involves several dimensions, including:

- Mission and strategy;
- Culture/structure and competencies (organizational and management values, management style, standards, organizational structure, core competencies);
- Processes (functions such as communication, planning, office management, relationships with other organizations, report writing, meeting facilitation, consensus building, research/policy development, monitoring and evaluation, performance management, financial and human resources management);
- Human resources (relationships with staff, members, management, external groups);
- Financial resources (both operating and capital required for the organization, fund raising, self-financing mechanisms);
- Information resources (media, electronic and paper resources management to support the mission and strategies of the organization);
- Infrastructure (physical assets, computer systems, telecommunications, productive work environments);
- Conflict management.

A major part of capacity development is at the *individual* level. This includes both members and non-members of fisher and other co-management organizations, as well as other beneficiaries of the programme. This is the most critical level as it involves the individual's capacity to function effectively and efficiently with the organization and, more broadly, with the co-management programme. Capacity development is designed according to the individual's function and relationship to the organization and the co-management programme.

Horton (2002) states:

It is often assumed that developing individual capacities will automatically lead to improved organizational capacity and performance. This is not the case. For example, there are many cases where individuals have developed skills in participatory research, but very few cases where participatory research has become institutionalized in the standard operating procedures of research or development organizations.

8.4.2. Approaches to capacity development

Capacity development cannot be 'done' by outsiders. An external agent can promote or stimulate capacity development and provide information, training and other types of support. But an external agent should not attempt to lead an organization's capacity development effort or take responsibility for it. The organization's managers and members must set goals and make decisions. Leadership must emerge from within the organization, and the organization's members must do most of the required work. However, an organization can benefit from external expertise and advice. But ultimately, the organization's own managers must be in the driver's seat (Horton, 2002).

Capacity development involves the acquisition of new knowledge and its application in the pursuit of individual and organizational goals. For this reason, learning by doing, or experimental learning, lies at the heart of capacity development (Horton, 2002).

The main tools for capacity development include one or more of the following approaches:

- Information dissemination;
- Training to develop knowledge, skills and attitudes;
- Facilitation and mentoring by an external agent;
- Networking, with the exchange of information and experiences from other people working on similar tasks, as well as through workshops, networks and communities of practice;
- Feedback, in order to promote learning from experience within an organization.

The type and amount of capacity development will depend upon the organizations' goals and the budget available for these activities. The provision of information or one-time training, while able to reach more individuals and organizations, seldom produces lasting changes in the participants' behaviour. Facilitation by an external agent is generally more effective, although it is more costly.

Capacity development is promoted by the following key factors:

- An external environment that is conducive to change;
- Top managers who are committed to provide leadership for change;
- A clear set of objectives and priorities;
- A critical mass of members involved in, and committed to, the change process;
- Awareness and understanding of the initiative;
- Open and transparent processes and decision-making;

- Adequate resources for developing capacities and implementing change;
- Adequate management of the capacity development process.

8.4.3. Strategic management

In planning and setting priorities for capacity development, there is a need to assess the factors that are limiting the organization's performance and identify those capacities that constrain performance the most. An organizational assessment is often needed to determine performance and capacity constraints and opportunities for change. Most organizations should undertake strategic management. Strategic management can be defined as an approach whereby an organization defines its overall character and mission, its longer term objectives and goals, the product/service it will provide, and the means (strategy) by which this can be achieved. Strategic management should address all dimensions of capacity at all three levels – individual, organization, system or enabling environment. This approach allows an organization to establish the desired relationship with other organizations or stakeholders within the broader co-management programme in which it functions. This requires a full and ongoing assessment of the strengths, weaknesses, opportunities and threats (SWOT) both externally and internally (Box 8.12). This approach should be participatory and consultative. Capacity assessments can be carried out as 'one off' types of initiatives, or they can be carried out at any one or all stages of co-management programme. For example, an organization embarking on involvement in co-management may need to develop initial capacities to carry out planning and negotiation, and other capacities later as the co-management programme matures. In all cases, emphasis would be given to utilizing existing capacities and to developing new capacities only as they are needed (UNDP, 1998).

8.4.4. Evaluation

Evaluation of the capacity development effort can serve two purposes. The first is accountability to determine if objectives have been achieved and resources used appropriately. The second is that evaluations are carried out to learn lessons that can be used to improve ongoing and future capacity development efforts (Horton, 2002).

An important aspect of capacity development is determining how the initiative can be sustained and remain relevant by responding to change. Sustainable change can be enhanced by:

- Involvement in decision-making;
- An atmosphere of mutual respect and trust;
- Building existing strengths and dimensions of capacity;
- Reward performance;
- Encourage new ideas and risk taking;

Box 8.12. SWOT Analysis.

SWOT stands for Strengths, Weaknesses, Opportunities and Threats. Strengths and weaknesses focus on the internal factors, while opportunities and threats reflect the influences of the external environment affecting the organization, community or activity. A strength is a positive attribute, a weakness is a negative attribute, opportunities are favourable factors in the environment, and threats are negative factors in the environment. The purpose of SWOT analysis is to identify strengths and opportunities and consider how to optimize them, and to identify weaknesses and threats and how these can be overcome. SWOT is a tool used in strategic management planning processes. It is used by an organization to assess its capability to become involved in co-management.
   Steps in a SWOT analysis include:

1. Clarify with the participants the specific item to be assessed.
2. Define terms in the context of the internal and external environments of the organization.
3. Ask participants to list strengths and post them under a column on a board labelled 'strengths'. Clarify and discuss the items listed. Group or cluster similar items.
4. Repeat the same process for weaknesses, opportunities and threats.
5. Analyse results. Ask the following questions:

• How can strengths be employed to take advantage of development opportunities or counteract them?
• How can weaknesses be overcome?
• How can the organization or activity maximize opportunities?
• How can threats be avoided?

6. Record responses and summarize major points. This will be the basis for identifying various actions and/or options.
7. Develop strategies and/or courses of action based on the responses.

   A matrix is produced that summarizes the key internal and external factors that influence the organization and various courses of action.

Source: IIRR (1998).

• Continual capacity development;
• Supportive leadership.

8.5. Social Communication

Successful change often depends on a variety of intangibles such as political will, trust, reputation and legitimacy. In addition, where suitable conditions exist or can be created, good communications can also be a key to bringing about change. Communications may be personal (one-to-one), inter-personal (among a few individuals) and social (when it involves social groups, such as a local community) (Borrini-Feyerabend *et al.*, 2000).

Social communication is the on-going flow of information and dialogue between the CO and the people in the community, and among the community members themselves about co-management, the programme, issues and solutions (Borrini-Feyerabend *et al.*, 2000). Effective social communication and information feedback mechanisms can empower and increase the involvement of community members in co-management. It provides for a better understanding of different stakeholders' perspectives and learning from different knowledge bases. For example, one objective of social communication can be to inform the community members about co-management and the need for co-management and to promote an open debate and understanding. Social communication can help community members gain ownership over the co-management programme. Borrini-Feyerabend *et al.* (2000) state:

> They (community members) do not merely aim at 'passing on a message about an issue' but at promoting its critical understanding and appropriation in society. After all, the most important result sought by a genuine co-management initiative is not for people to 'behave' in tune with what some experts believe it is right for them, but for people to think, find agreements and act together on their own accord.
>
> (p. 27)

The development of a social communication initiative should begin by understanding how people in the community communicate and discuss ideas and issues among themselves. It involves understanding the system of local media. This may range from word of mouth at the fish landing site or market, to call-in radio talk shows, to songs, to more formal gatherings. Social communication can take a variety of different forms and creativity is important to be effective. These forms range from one-to-one dialogue and group meetings to the use of mass media such as radio, television and Internet. It is important to avoid using a 'teaching' or 'preaching' attitude and instead to promote dialogue and open discussion of different points of view (Borrini-Feyerabend *et al.*, 2000).

Social communication may entail different forms of communication such as informing, raising awareness and training. While these forms of communication can be interactive between the sender and receiver, they are generally carried out in a 'top-down' manner with the sender in control of the message. Interactive learning is a different form of communication which enhances common knowledge, awareness and skills by having those with different views think, discuss and act together (Borrini-Feyerabend *et al.*, 2000).

The knowledge of how to communicate, the ability to communicate, and feeling comfortable in communicating are all skills which may need to be developed. Training in communication skills (both talking and listening) may need to be a precursor to a social communication programme. Training may need to target different groups with different communication needs, such as government officials and staff and community members.

8.5.1. Social marketing

One approach to social communication is social marketing. Social marketing is the use of the philosophy, tools and practices of commercial marketing to 'sell' a social cause, idea or behaviour change to a targeted group of individuals (Boxes 8.13 and 8.14). Social marketing can be used to accept a new behaviour, reject a potential behaviour, modify a current behaviour and/or abandon an old behaviour. The core unit of social marketing is the exchange between at least two parties. For example, the CO provides information to the target audience and the hoped for exchange is a change in behaviour, such as elimination of destructive fishing practices. The focus of social marketing is the community members. It is critical to think you know what is going on in their heads but to really find out the why and what.

The social marketing approach is a process that is very systematic and planned, with all stages of the programme clearly mapped out with objectives and behavioural targets. Steps in the process are:

- Define the problem based on analysis and community assessment;
- Identify behavioural change/actions that could eliminate/reduce the problem;
- Identify potential audience segment/target;
- Conduct root cause analysis – just ask 'why?';
- Establish goals and objectives for the programme;
- Position the intervention: product, price, place, promotion and politics;
- Deliver and monitor the intervention;
- Evaluate the programme (Kolter *et al.*, 2002).

Among the more effective tools used in social marketing are advertising and public relations, promotions and publicity via mass media, special events,

---

Box 8.13. Mobilizing the Private Sector for Action, 'Our Seas, Our Life' in the Philippines.

1998 was the International Year of the Ocean (IYO). The Coastal Resources Management Programme, a US Agency for International Development-funded programme for coastal management in the Philippines, in partnership with the National Commission on Marine Sciences with support from Silliman University, National Museum, and the Department of Environment and Natural Resources Protected Areas and Wildlife Bureau and a host of private sector sponsors organized the 'Our Seas, Our Life' travelling exhibit. The exhibit was launched in Cebu City in February 1998 and travelled to key cities in the Philippines until December 1999, drawing approximately 1.4 million viewers. A huge success, the exhibit proved invaluable in calling national media and public attention to coastal issues. It was also a highly effective social marketing tool, providing a forum for discussion of coastal resource management problems and solutions among a wide range of sectors in the cities visited.

Source: DENR *et al.* (2001b).

Box 8.14. Building Support for Coral Reef Management in Phuket, Thailand.

Phuket, a rapidly growing tourism area in southern Thailand, has a significant coral reef area that provides economic benefits to fishers and tourism. The Phuket Coral Protection Strategy, a project implemented by the University of Rhode Island and the Phuket Marine Biological Center of the Department of Fisheries, had the goals to protect and provide for the sustainable use of the reefs and to build local and national support for coral reef management.

Throughout the year and a half of the issue identification and analysis stage, considerable effort was made to heighten public awareness of coral reefs and to build support for subsequent management initiatives. At the time of the project there was virtually no awareness of coral reefs and their importance to the economy. Early activities, which included media campaigns, community events and the publication of brochures, were designed to enhance both the general public's and the private sector's appreciation for the area's reefs and to explain why a protection strategy was necessary. Support for coral protection was also built through the extensive discussions carried out with reef-dependent businesses and reef users during the process of issue identification.

Source: Hale and Lemay (1994).

---

celebrity endorsements, testimonials, and advocacy campaigns. For example, a slogan or vision statement, such as 'Our Seas, Our Future!', can be adopted to convey a message and develop solidarity. Any slogan should evolve with community members rather than be chosen by outsiders.

8.5.2. Communication

Social communication can be used to mobilize community members to participate in co-management. Participatory techniques can be used to raise people's awareness, knowledge, ability and motivation to make decisions about their future. Common participatory techniques include workshops, public meetings, study tours, community exchange visits, advocacy campaigns, debates, street theatre, committees, community patrols, citizen watchdog groups, school programmes and special projects involving the community or various sectors of society (DENR *et al.*, 2001b).

Social communication should promote internal discussion within different stakeholder groups and organizations. Discussion allows different viewpoints to be aired and discussed, trust and credibility to be created, and group cohesion to be strengthened. This can be accomplished by building on a common focus or issue and holding meetings that foster contact and trust and allow bridges to be built among members.

Social communication should be respectful of local cultural traits and norms. Any information conveyed should be truthful, fair and reasonably complete (Borrini-Feyerabend *et al.*, 2000). Social communication should be ongoing throughout the life of the project and beyond.

9 Community Organizing

J. Parks.

The active participation of people in a community in the co-management programme is at the heart of co-management. Success of co-management is directly related to a well-organized community that has been empowered to take action to manage and conserve its aquatic resources. Community organizing is much more than just establishing organizations, it is a process of empowerment, building awareness, promoting new values and behaviours, establishing self-reliance, building relationships, developing organizations and leadership, and enabling communities to take action (Table 9.1). Thus, as mentioned above, environmental education, capacity development and social communication are central elements of the co-management process.

Table 9.1. Community organizing.

Stakeholder	Role
Fishers	• Participation in meetings • Support organization formation • Assist in developing organization structure • Support and participate in organization • Leadership
Fisher leaders/core group	• Participate in meetings • Assess situation • Decide on mission of organization • Canvas community for support • Support consensus process • Develop organizational structure
Government	• Support organizing efforts • Provide legal support to organize
External agent/CO	• Identify core leaders • Organize core group • Support core group in mobilization • Build alliances and networking • Seek funding for organization

Community organizing looks at collective solutions. It changes the balance of power and creates new power bases. It is a value-based process by which people are brought together in organizations to jointly act in the interest of their 'communities' and the common good.

It has been reported that the fundamental source of cohesion of every strong community organization is the conviction that it offers its members a unique vehicle for exercising and developing their capacities as citizens. The empowerment process at the heart of community organizing promotes participation of people, organizations and communities towards the goals of increased individual and community control, political efficacy, improved quality of community life and social justice (Wallerstein, 1992).

To participate in co-management, the stakeholders will need to organize themselves and arrive at an internal consensus on the interests and concerns that they want brought forward. Meetings and discussions are held among the individual stakeholders to identify and clarify their interests and concerns and for those individuals with common interests and concerns to organize themselves into groups. Effective community participation in co-management requires a strong community organization(s) to represent its members (Box 9.1). In some cases, community organizations capable of representing their members in co-management already exist in the community. In other cases, organizations will either need to be strengthened or newly established. One or more community organizations may be needed in the community depending upon its size, diversity and needs. An appropriate person(s) from the organization must be selected to represent them on the larger co-management organization.

Box 9.1. Fisher Organizations in the Caribbean.

The Caribbean has fewer community organizations that are positioned to play roles in co-management than in other regions of the world. The low level of experience with collective action in the region is evident in that fisher organizations are not widespread, are relatively weak, and are not well prepared for a role in co-management. Community organizing will be a critical component of introducing or strengthening co-management in the Caribbean. Weaknesses of fishery organizations in the Caribbean suggest that much will have to be done to promote sustained collective action to institutionalize co-management. Crisis-driven management responses prevail in both government and industry, and crisis responses often feature intense, temporary collective action. Some countries have cooperatives and fisherfolk organizations. However, these groups will not automatically be suitable as representative organizations in co-management. It is likely that they were established with objectives that relate more to expanding exploitation, improving marketing and increasing incomes of members. Changes in outlook will be necessary for these groups to play major roles in resource management. These changes will be difficult and lengthy, especially if the organization is still struggling with its original mandate. Putting more focus on resource management may strain the internal cohesion of the organization.

Source: McConney *et al.* (2003b).

Fishing cooperatives and fisher associations exist in many communities. However, these organizations will not automatically be suitable as representative organizations in co-management. It is likely that they were established with objectives that relate more to expanding exploitation, improving marketing and increasing the incomes of members. Changes in outlook will be necessary for these organizations to play major roles in resource management. These changes may be difficult and lengthy, especially if the organization is still struggling with its original mandate. Putting more focus on management may strain the internal cohesion of the organization (McConney *et al.*, 2003b).

The process of community organizing is seldom 'tidy'; it doesn't always happen in neat, predictable steps. Activities may occur simultaneously. Community organizing involves learning, sharing and adapting. It often involves building upon existing institutions and organizations in the community. Community organizing is led by the CO, but must be a collaborative effort of all the stakeholders in co-management.

9.1. Components of Community Organizing

There are several components in community organizing, some of which have been undertaken earlier in the process:

1. Preparation
• Create a core group(s) and core leaders;

- Assess the situation (research);
- Hold visioning exercises;
- Decide on a mission for the organization.

2. Mobilization (Boxes 9.2 and 9.3)

- Seek out community support and build a base of support among community members;
- Hold meeting(s) to discuss the vision or mission, reach consensus and agree on developing an organization or join an existing organization;
- Develop organizational goals and objectives, organizational structure, leadership/membership and action plan;
- Appoint a representative of the organization.

Box 9.2. Community Organizing in the Fisheries Sector Programme in the Philippines.

To become managers of coastal resources, fisherfolk in each village were guided to form an association which was eventually registered with the Cooperative Development Authority as a cooperative. The process began with the formation of a core group which recruited more members and initiated the formulation of the association's articles of cooperation and by-laws. The expanded membership ratified the articles and by-laws and elected officials who, with help from the NGO, prepared and submitted the required registration papers.

Training on the knowledge, skills and attitudes required for cooperative management included leadership, team-building and organizational skills; participatory development planning; simplified financial and accounting systems; project feasibility studies; and problem solving and decision-making. Training in fisheries-related laws and ordinances and in the knowledge and skills for managing fish sanctuaries, mangroves and artificial reefs was also provided.

Source: Abad (1997).

---

3. Strengthening (Box 9.4)

- Environmental education, capacity development and social communication;
- Building alliances and networking;
- Organizational sustainability to keep members and funding.

4. Evaluation.

9.2. Preparation

Community organizing starts with the preparatory component which involves the CO working with a leader(s) from the community to establish a core group

Box 9.3. The Community Organizing Process in Peam Krasaop Wildlife Sanctuary (PKWS), Koh Kong Province, Cambodia.

In Cambodia, the word 'community' has come to mean a special group of people interested in organizing themselves and supported by programmes of NGOs or government agencies. For example, Cambodians refer to such things as community forestry, community fisheries, community land use planning, and community protected area, where these communities have been supported by government and NGOs. Community, in this sense, may refer to a committee in a village, which is not necessarily the administrative boundary of a village or commune.

In Peam Krasaop Wildlife Sanctuary (PKWS), Koh Kong Province, the Participatory Management of Mangrove Resources (PMMR) has been doing community organizing work since 1997 to facilitate a process where a village structure called the Village Management Committee (VMC) is set up and strengthened. PMMR is based at the Ministry of Environment. The project team includes technical staff from the Ministry at the national level, and staff from technical line departments in the province such as: the Department of Environment, the Department of Rural Development and the Department of Agriculture, Forestry and Fishery.

When PMMR began this process in PKWS, meetings were held to discuss the community organizing approach. The PMMR project arranged a series of workshops and study tours in and outside the country for the key villagers and the PMMR team members. Inside the country, study tours were conducted to similar projects of community-based natural resources management, such as Community Fisheries at Ream National Park, Sihanouk Ville and Community Forestry, FAO Project in Siem Reap province. Trips outside of Cambodia were held to see the participation of local community in coastal zone management in Thailand and Sri Lanka. These study tours provided challenges for the participants to exchange ideas and learn experiences from one another. Learning from training/workshops and study tours, the community members expressed an interest in working with the PMMR team and concerned government agencies to come up with a community management strategy that would work in this area.

In planning for the community organizing process, PMMR suggests considering the following questions:

- What are the villagers' concepts/ideas for protecting their fishing grounds?
- Are the identified fishing grounds to be co-managed more or less free from interruptions by outsiders?
- Should the boundary for the community-managed area follow an administrative boundary or a natural physical boundary?
- What are the steps in establishing community fishing areas?
- Who is involved in the process of community regulation development?
- How does one gain official recognition of community regulations from local authorities and technical departments?
- What will community by-law look like? For example, will penalties be included, what sizes of gear will be restricted?

In the PMMR's experience, community organizing is both a technique for problem solving and a way to improve income for people, strengthen local awareness and enhance natural environment. Communities are enabled to consider their problems on economic, political and social needs, and initiate conflict resolution within a natural resources management framework. In the community organizing process, local community capacity building is a central point. The community development worker or field facilitator must understand well any issues and factors that affect villagers' participation in the community organizing process.

In Koh Kong, four communities were established with strong support by the Provincial Governor and the Minister of Environment combined with good facilitation by the PMMR research team, and learning from other experiences.

Source: Nong *et al.* (2004).

Box 9.4. Local Government–Civil Society Group Partnership, Philippines.

During the past 14 years, Naga City, Bicol Province, Philippines has built a reputation for being a model local government unit that pioneered innovations in local governance. For example, the People Empowerment Programme (PEP) is a continuing initiative of the city government to promote political empowerment of its citizenry. In particular, the city government passed an 'Empowerment Ordinance' that mandated the partnership between the local government and the non-government sectors in the city and encouraged the federation of the NGOs and people's organizations into the Naga City People's Council (NCPC). The NCPC continues to improve the city's conditions, with funding and technical assistance from the city government, the USAID-funded Governance and Local Democracy Project and, more recently, the Philippine–Australia Governance Facility for institutional development and the strengthening of the city's basic sectors.

By giving premium to community organizing, the NCPC has given marginalized sectors a voice and an avenue for meaningful participation in governing their city. The NCPC even extended its reach by organizing a functional Barangay People's Council. In doing so, participation and inclusiveness in direction-setting, policy-making, as well as programme and project implementation, monitoring and evaluation at the city level, has been widened. In addition, there is now a heightened level of trust, confidence and openness between civil society groups and government.

The innovative approach of the Naga City government now includes multi-stakeholder planning for resource management and rehabilitation. Under the Naga City Participatory Planning Initiatives, the rehabilitation of Naga River has started. So far, a draft of the Naga City River Watershed Strategic Management Plan was produced based on extensive inputs from stakeholders who have approved the plan themselves through a stakeholder congress.

Source: www.naga.gov.ph

---

to lead the community organization. This core group should be small (three to five people) and include individuals who are enthusiastic, have a common interest, and represent a cross-section of the community. The leaders should be individuals who are acceptable to the community and who command sufficient respect.

The core group should hold meetings to assess the situation in the area. Information from the research activities conducted earlier can be utilized to gain a broad understanding of the situation. People in the community may be asked to identify issues.

The core group then decides on the initial direction of the organization and a mission for the organization. Questions to be asked include:

- What are we trying to do?
- What size of area are we going to organize?
- Who will support our efforts?
- What is a good idea for our first action?
- How are we going to reach out to others?

9.2.1. Leadership

Any community organization rises or falls with the quality of its leadership. The importance of identifying and developing responsive and effective leadership from the community cannot be understated. Leaders are needed to direct change and mobilize people towards a common vision. Leaders are not necessarily born with special qualities. Leadership can be learned and practised. Leadership and management work hand-in-hand and are sometimes done by the same person, but they are not the same thing. A manager runs the day-to-day operation of the organization. A leader provides inspiration and motivation. Every member should be encouraged to take leadership roles. Leaders should be changed regularly to discourage corruption. Future leaders should be identified and trained on a continuous basis (Box 9.5).

It is never easy to find community members who have the time to devote to organizational tasks. Often, the potential leaders with more time on their hands are from economically well-off families. The poorer ones are often too busy earning their living. It is important for a CO to look for potential leaders among the poorer sectors in the community who could provide even a little of their time to the organization. Equally important is to always be on the look out for possible women leaders. Not too many women are visible in community affairs but developing potential leaders among them is an important element in a co-management process.

Box 9.5. Local Leaders in Bangladesh.

In Bangladesh, local leaders of the *baors* (inland water bodies) were identified and elected by the fishers. Leaders' terms of office were limited so as to give others the chance to gain leadership skills and to reduce the possibility of corruption. Reliance on one individual as a leader can be a problem. In certain Philippine cases, projects failed when the leader died, left political office, or left the area, because there was no one to take the leader's place.

Source: Pomeroy *et al.* (2001).

Thus, leaders should represent, to the extent possible, the varied interests in the community. In some cases, co-management initiatives start with leaders from a specific sector or group in the community but this should always be expanded as the process moves along to ensure broader participation. In community organizing, first- and second-line leaders are also developed among community leaders with the former mentoring the latter.

Effective leaders:

- Challenge the process (pioneers, search for opportunities, experiment, take risks);
- Inspire a shared vision (visionaries, envision the future, enlist others);
- Model the way (practise what they preach, set an example, plan small wins);
- Enable others to act (team players, foster collaboration, strengthen others);
- Encourage the heart (coaches and cheerleaders, recognize contributions, celebrate accomplishments);
- Welcome criticism.

9.3. Mobilization

After the core group has decided upon a mission statement for the organization, they canvas the community for support and build a base of support among community members. The community members are contacted one-on-one by the core group members and CO to request their opinions/advice on the identified mission, including activities they themselves can undertake. The individuals should be encouraged to discuss their concerns and the costs and benefits it could bring to them and the community. This process provides an opportunity to gain insight into the perceptions and interests of stakeholders and to identify common interests and potential conflicts. It also allows individuals with common interests to be brought together. Non-organized fishers can be identified and asked about becoming members of an organization. Go to where people are. Try to include those who are under-represented.

Meeting(s) are held to discuss the purpose/objective, reach consensus, and to agree on developing an organization or join an existing organization. The core group organizes a series of meetings with stakeholders to discuss the mission and to share their views. The first meeting should be started on a neutral tone and do not deal with sensitive topics initially. The first meeting must focus on the concerns of the fishers and the possible need for them to form an organization to address these needs. The main objective of the meeting should be to reach agreement about the formation of an organization.

Encourage people to come together when they are ready and not when it is imposed on them. Be aware of power structures within communities and institutions which may inhibit some stakeholders from contributing. The facilitation of these meetings is crucial to their success. If meetings are well-managed, they can provide an opportunity for each stakeholder to hear and appreciate others' views and concerns. Social communication activities can be used to help ensure support for the organization and its mission.

Every stakeholder will have different information, concerns and interests which need to be considered and developed. Making sure that all stakeholders are able to develop their own position and form of representation may initially result in challenges to community organizing. It must be remembered that building an organization is a slow process. People need to feel that being part of an organized group is necessary to protect their interests.

Once there is agreement on the organization, a meeting(s) is held to specifically define the organization's goals and objectives, organizational structure, leadership, membership, dues and finances and action plan. Organization structure is the framework around which the group is organized. Structure describes how members are accepted, how leadership is chosen, and how decisions are made. Structure give members clear guidelines for how to proceed and it binds members together. There are three elements to organizational structure:

1. Some kind of governance to make decisions;
2. Rules by which the organization operates; and
3. A distribution of work.

Each organization may develop roles for individuals to play in the organization. There may be a variety of committees within an organization, such as the executive committee, action committee and finance committee, which carry out specific roles and responsibilities. Organizational structure is best decided upon internally, through a process of critical thinking and discussion. Organizational structure will be guided by such factors as the purpose of the organization, size, volunteer or paid staff, and whether it is advocacy- or service-oriented (Boxes 9.6 and 9.7).

Successful organizational structure usually includes:

• An elected leadership;
• Regular meetings;
• A newsletter;
• A means of delegating tasks and responsibilities;
• Training for new members;
• Social time together;
• A planning process;
• Working relationships with power players and resource organizations.

A goal is set and objectives are devised that will lead to the goal. To be effective the organization should, at least initially, pursue only one objective at a time. The organization should generate ideas to achieve the objective, then carry them forward into an action. Once the organization's members agree on an action, create an action plan. The action plan should identify strategies to achieve each objective. Work should be broken down into manageable tasks. The action plan should include a timeline to identify when things should be done and by whom, an ordered list of tasks to complete, persons responsible for each task, facilities and funds.

Box 9.6. Beach Village Committees in Malawi.

The Participatory Fisheries Management Programme (PFMP) on Lake Malombe and the Upper Shire River in Malawi was implemented to reverse the decline in the fisheries. The strategy employed to implement the PFMP involved creation of a Community Liaison Unit (CLU) composed of fisheries extension staff and Beach Village Committees (BVCs) representing the fishing communities. The BVCs were intended to serve as the basis for a two-way channel of communication between fishers and the Fisheries Department. It was also hoped that they would progressively assume responsibility for the management of the fishery. The BVCs were selected by village communities. The BVC was composed of gear owners, fishing crew members, and any active member of the village group. The village head was supposed to serve as an advisor to the BVC. The BVC had the role of controlling the beach and the group that fished from the beach, limit access to the beach and fishery, organize meetings of its members, establish fishing rules and represent its members at higher levels.

A number of problems occurred in the establishment and operation of the BVCs. Some local leaders appointed certain individuals for their personal gain. Most BVCs had very few fishers as members and even fewer had crew members.

The creation of BVCs resulted in contests for power and authority with the village headmen. In many cases, the village headman became the *de facto* leader of the BVC, even though he was not the chairman. The village headmen became prone to ignoring the authority of the BVCs, since by historical tradition and custom, they hold ultimate authority. Where the BVCs have resisted being taken over by village headmen and have established some semblance of independent authority, fishers have often been confronted with dual authority of both the BVC and the village headman. In some cases, gear owners saw this as a chance to cut out the practice of giving payment to the village headman and to challenge them on this matter. The strong village headmen had no option but to curb the powers of the BVCs. In some cases, the elected BVCs were forced to disband and were replaced by BVCs with members appointed by the headman. In these cases, fishers saw the BVCs representing the village headman's interests more than their own.

In contrast to Lake Malombe and the Upper Shire River, the village headmen were largely kept out of the BVCs on Lake Chuita in Malawi. Since the fishers elected BVCs on their own with the facilitation of the Fisheries Department, they were able to ignore the Lake Malombe model which incorporated village headmen as ex-officio members. The exclusion of headmen from the BVCs and the growing reluctance of fishers to make payments to the headmen have increased tension between the BVCs and the headmen.

Source: Njaya (2002).

Members and leaders make all organizational decisions, from by-laws to slogans. Members raise and select organizational issues based on the self-interest of the group, and broad agreement among members is necessary before the organization will pursue an issue.

Each organization should discuss, agree on, and post guidelines for decision-making. Some decision-making approaches include straw polling (a show of hands to see how the group feels about an issue), voting and consensus (bringing the group to mutual agreement by addressing all concerns).

Box 9.7. Barbados National Union of Fisherfolk Organizations.

The Barbados National Union of Fisherfolk Organizations (BARNUFO) is a secondary, or umbrella, fishing industry organization. It is not a trade union, but an alliance or federation. BARNUFO's mission, according to the written constitution, is to fulfil the requirements of its member fisherfolk organizations with a view to improving their socio-economic conditions based on sustainable development of fisheries 'from the hook to the cook'. The members of BARNUFO are the primary fisherfolk organizations of Barbados, not the individuals in the industry. Two persons can be selected from each primary member organization to be representatives in BARNUFO. BARNUFO sits on the government's Fishery Advisory Committee. The Fisheries Division works with BARNUFO as its main partner in sea egg (sea urchin) co-management initiatives such as data collection.

Source: McConney *et al.* (2003a).

The organization, if they want to take part in the co-management negotiation process, will need to identify and appoint one or more individuals to represent them. In cohesive organizations this may be easier than in non-cohesive organizations. Criteria such as honesty, knowledge of the area and issues, negotiation skills, maturity and status in the community and others may be used to select a representative.

9.4. Strengthening

To ensure sustainability of the organization, there must be continual strengthening of the organization and its members; this includes government. Strengthening means obtaining the necessary attitudes, knowledge, skills and resources to take part in the co-management programme. This can be achieved through the environmental education, capacity development and social communication activities described in Chapter 8. Change does not occur in the blink of an eye. Individual change in terms of values and awareness takes time to change and must be strengthened through information, education and capacity development. The value transformation of individuals is enhanced through an organization resulting in collective action.

In addition, the CO should help the organization in building strategic alliances with other organizations with common interests. Networking establishes linkages with other organizations working for a common goal. The strategy is to share information with other organizations so as to bring about greater understanding as well as social and policy change.

Once the organization is formed, efforts need to be made for sustainability. Several measures can be undertaken (Espeut, nd):

• Members should be notified of meetings in good time.
• One of the most difficult pitfalls in community organizations has to do with

keeping up the momentum. Almost inevitably, the attendance at meetings will be high initially and then fall as the meetings become dull and routine. Each meeting should offer some particular issue for discussion which will captivate the members, or there can be a special person invited to address the meeting, or there can be a fisheries education exercise.

- Personality clashes can weaken or even be the death of community organizations, as very few people have the time or the energy to come to meetings which are nothing more than a collection of individual ego trips. For community organizations to be strengthened, ways have to be found to neutralize these personality problems. The problem people need to be spoken to and shown how their behaviour is affecting others. Special training may be needed to be able to handle these cases.
- Financial irregularities can affect any organization. To address this problem, there should be suitable training on financial accounting, there should be frequent financial reports to the members, and there should be a bank account where all organization funds are deposited.
- Meetings should be an opportunity for informing the members and community about what is happening and planned. Any hint of secrecy should be absent from a well-run organization.
- Stay in touch with each other. Every opportunity should be taken to provide information to members, such as a newsletter. Representatives of the organization to the larger co-management organization should report to the membership not just the leadership.
- Act more, meet less.
- Keep time demands modest.
- Provide social time and activities.
- Work in pairs to improve communication, make work less lonely and ensure tasks get done.
- Provide skills training.

9.5. Evaluation

Evaluation is used to measure the success or failure of the organization to meet its goals and objectives and the reasons why. Monitoring is a continuous process of gathering information about the organization and its management. The membership of the organization, as well as community members, should be involved in the monitoring and evaluation. Evaluation should be an iterative process which leads to modification and improvement of the organization to better suit the needs of its members and current conditions. After each activity has been implemented and completed, the evaluation should ask:

- What has been accomplished?
- What still needs to be done?
- What was done well?
- What could have been done better?

10 Co-management Plan and Agreement

D. Wilson.

The community profile provides information on the resource, social, economic, cultural and institutional/legal conditions in the community. The community profile identifies key issues to be addressed by a fisheries and coastal management plan. The preparation and adoption of the co-management plan and agreement requires the involvement and support of all stakeholders. This is provided by community organizations with members who are empowered to participate in the co-management programme (DENR *et al.*, 2001b).

The co-management plan provides a way of organizing all stakeholders in order to prioritize issues, define policies, implementation and make more informed decisions for the sustainable use and management of fisheries

© International Development Research Centre 2006. *Fishery Co-management: A Practical Handbook* (R.S. Pomeroy and R. Rivera-Guieb)

Table 10.1. Co-management plan and agreement.

Stakeholder	Role
Fisher/fisher organization	• Participation in negotiation and planning • Provide input into formulation of goals and objectives • Provide information and feedback on plan • Mission and vision setting • Conduct small group discussions • Participate in co-management organization • Build community consensus
Other stakeholders	• Participation in negotiation and planning • Provide information and feedback on process
Government	• Provide basic policies, legal and planning framework • Participate in negotiation and planning activities • Assist in identifying revenue sources • Participate in co-management organization • Conduct community meeting • Clarify implementing responsibilities
External agent/CO	• Facilitate process of negotiation and planning • Provide technical assistance and training • Ensure participation of organizations and other stakeholders • Convene stakeholders and groups for planning • Assist in establishing co-management organization • Train in negotiation and planning processes • Assist in mission and vision setting • Build consensus

resources (Table 10.1). The co-management planning process involves several activities including:

- Establishing a co-management body;
- Agreeing on rules and procedures for negotiation;
- Developing a mission statement;
- Establishing a management unit;
- Negotiating co-management plans and agreements;
- Plan – goals, objectives and activities;
- Monitoring and evaluation plan;
- Developing a co-management agreement;
- Establish a co-management organization;
- Revenue generation and self-financing;
- Legal support;
- Publicizing.

Negotiation is at the heart of the co-management process. Any plan or agreement is only as good as the process that generated it. The challenge for the planning process is to develop a partnership among the stakeholders in which there is sharing and negotiation undertaken in an efficient and equitable manner (Borrini-Feyerabend *et al.*, 2000).

Borrini-Feyerabend *et al.* (2000) have identified several important points to be remembered:

- There are many management options, good and bad;
- Given the complexity of ecological and social systems in fisheries, the best approach is adaptive management ('learning-by-doing');
- Conflicts of interest between the stakeholders are inevitable but can be managed, and all the more if recognized early on;
- Even when a satisfactory management solution has been found, it will not remain valid forever; conditions will change and the solutions will have to change in response;
- All the stakeholders (especially the professional experts) should adopt a mature, non-paternalistic attitude, and acknowledge the legitimacy of interests and opinions different from their own; and
- All stakeholders should be given a fair means to have their claims and issues heard and discussed.

The co-management plan includes explicit identification of management strategies and actions, as well as co-management roles and responsibilities among the partners. The co-management plan should be structured around the key components of capability building, resource management, community and economic development, livelihood development and institutional development. The co-management plan should reflect the community's vision for the future. Initially, the plan may be focused on a specific issue or problem, such as reducing illegal fishing; but in better prepared communities or in more mature stages of the programme, the co-management plan may cover several issues or problems in a more integrated manner.

In developing the goal, objectives and activities, transparency is important if the management system is to be a learning one. This is particularly important for a process with uncertain outcomes. Therefore, documentation of process, plan, and outputs of and inputs to the process are necessary for the system to work. Transparency is also important if stakeholders are to be properly informed so they can participate fully. For some stakeholders, the process must go further, educating them to prepare for participation.

10.1. Adaptive Management

As presented in this handbook, the co-management plan is based on an approach called adaptive management. Adaptive management is 'the integration of design, management and monitoring to systematically test assumptions in order to adapt and learn' (Margoluis and Salafsky, 1998). It is 'an approach based on the recognition that the management of natural resources is always experimental, that we can learn from implemented activities, and that natural resource management can be improved on the basis of what has been learned' (Borrini-Feyerabend *et al.*, 2000). Adaptive management supports the need to develop plans and agreements that can be renegotiated to meet changing needs and conditions (Box 10.1).

> **Box 10.1.** Adaptive Management.
>
> All good fishers learn from their successes and failures. A fisher will try a new fishing method, for example, monitor the results, and see how the results compare to what was predicted to happen. Based on the new information, the fisher may accept the fishing method, may adapt the fishing method to improve on it, or may reject it. This learning and adaptation is the basis for adaptive management. Adaptive management goes one step further and relies on systematic feedback learning and the progressive accumulation of knowledge for improved fisheries management. Adaptive management relies on deliberate experimentation followed by systematic monitoring of the results from which the fisheries managers and fishers can learn. Adaptive management is participatory, involving fishers as partners with fisheries managers in the management process.

Adaptive management takes the view that fisheries management can be treated as 'experiments' from which managers and fishers can learn. Adaptive management differs from the conventional practice of fisheries management by emphasizing the importance of feedback from the fishery in shaping policy, followed by further systematic experimentation to shape subsequent policy, and so on. In other words, it is iterative, repeating a process of steps to bring the manager and fisher closer to a desired result. Each iteration should involve making progress in reaching established goals and objectives. The important point is that effective learning occurs not only on the basis of management success but also failures. However, learning from failures presupposes that what is learned can also be remembered. Organizations and institutions can learn as individuals do, and adaptive management is based on social and institutional learning. The mechanism for institutional learning involves documenting decisions, evaluating results and responding to evaluation. Institutional learning must be embedded in both fisheries managers and the fishers, and the knowledge held by each must be respected and shared.

For example, of particular importance are environmental fluctuations. Many areas are seeing decadal-scale shifts in marine ecosystems, as well as large, infrequent disturbances. How can we best respond to such changes? Some government agencies keep records or maintain disaster-response plans. But fishers themselves maintain institutional memory of such fluctuations, along with response mechanisms. Perhaps a combination of agency institutional memory and fishers' knowledge can help provide adaptive responses for ecosystem changes.

The success of adaptive management will depend on fisheries managers and fishers keeping an open mind to work together and share knowledge. Recognition that non-scientific forms of knowledge may nevertheless be valid and important can promote understanding in a genuinely participatory manner (Box 10.2).

The adaptive management framework involves first thinking about the situation in the fishery, collecting information about the fishery, and developing a specific assumption about how a given intervention will achieve

---

Box 10.2. Using Adaptive Management in Fisheries.

ADMAFISH is a continually evolving programme that actively applies the principles of adaptive management to identify, test and refine methodologies that support co-management. ADMAFISH monitors fish catch throughout the project area primarily through village monitors. These village monitors, retired and active fishers, collect catch data from the fishers at the landing site and ensure that they follow the rules. Data collected through these efforts are used to ensure that proper fishing fees are paid to the co-management organization. They are also used to continually adjust the fishing quotas set by the co-management organization. If there is too much catch of a certain species, then the quota is adjusted down. This investment in data collection and analysis has resulted in a co-management programme that allows for fisher input, questions about the resource to be asked and answered, and rules to be enforced.

---

a desired outcome. The intervention is implemented and the actual results are monitored to determine how they compare to the ones predicted by the assumptions. The key is to develop an understanding of not only which interventions work and which do not, but also why.

Adaptation is about systematically using the results of the monitoring to improve the intervention. If the intervention did not achieve the expected results, it is because either the assumptions were wrong, the interventions were poorly executed, the conditions at the intervention site had changed, the monitoring was faulty, or some combination of these problems. Adaptation involves changing the assumptions and the interventions to respond to new information obtained through the monitoring efforts.

Finally, learning is about systematically documenting the process that was followed and the results that were achieved. This documentation will help to avoid making mistakes in the future.

10.2. Preparing for the Planning Process

(The source for much of this section is: Borrini-Feyerabend *et al.*, 2000.)

Prior to beginning the negotiation, rules, procedures and equity issues need to be considered. Any negotiation is based on cultural and political realities. There may already exist in the community procedures for developing and negotiating plans and agreements that are simple, effective and inexpensive. These procedures may or may not be suitable for the co-management planning process. In cases where they are not suitable or where no procedures exist, new procedures will need to be designed. The CO may propose a schedule of meetings, some rules and procedures for participation, and a structure for a co-management body to conduct and oversee the planning process. The CO should present this proposal to representatives from the local government and community organizations. They may discuss and modify them, but a first meeting among the co-management stakeholders is proposed. The CO should work with the stakeholders to obtain an agreement on the place, date, hour, participants, agenda, logistics and facilities needed for the meeting that will begin the actual co-management process. It is important to consider fairness and equity issues; that is, to ensure that the previous less empowered segments of the community are fully involved in the process (Boxes 10.3 and 10.4). This is done through the community organizations, but also by recognizing entitlements and promoting fair negotiations (Borrini-Feyerabend *et al.*, 2000).

Edmunds and Wollenberg (2002) have identified a number of steps for negotiation, especially for disadvantaged groups:

- Neutral conditions are created including time, place, language, terms of communication, materials for illiterate participants and information sources.
- Inform participants fully about to whom conveners and facilitators are accountable.
- Create possibilities for disadvantaged groups to pursue alliances with more powerful groups in negotiations.
- Acknowledge the right of disadvantaged groups to identify 'non-negotiable' topics or items they view as inappropriate for discussion in the negotiation.

Box 10.3. Issues in Plan Preparation – Differing Objectives and Co-management Arrangements.

Two issues are critical to consider and address while in the co-management plan preparation process since the issues can cause problems later in the programme. The first is the analysis of stakeholder objectives for wanting to engage in co-management. Each individual stakeholder and stakeholder group, including government and community organization, has their own reasons and objectives to engage in co-management. These may be similar or may be widely different. During the co-management plan process these objectives need to be openly and clearly stated, clarified, discussed and negotiated among the stakeholders. If they are not, it can cause problems later as the stakeholders will have differing expectations and takes on why they are involved in the programme.

The second issue is where on the continuum of co-management arrangements to begin the co-management programme. As presented in Chapter 2, Section 2.1, there is a hierarchy or range of possible co-management arrangements, from those in which government consults fishers before making a decision, to those in which the community of fishers make decisions and consult with government. It is generally acknowledged that not all responsibility and authority should be vested at the community level. The amount and type of responsibility and/or authority that the partners have will need to be negotiated during the planning stage. It will depend on legal authority, government's willingness to devolve responsibility and authority, the capacity of the partners, among other issues. Too much responsibility and authority being retained by government may stifle community initiative, while too much responsibility and authority too early for the community may lead to problems due to lack of capacity. The partners need to negotiate the initial arrangement from positions of equal power, which is why organizing and empowerment activities are undertaken early in the co-management programme, and be realistic about what they are capable of doing.

Box 10.4. Promoting Equity in Co-management: Some Examples and Ideas.

- Disseminating information on environmental values, opportunities and risks of relevance to potential stakeholders.
- Disseminating information on various natural resource management options.
- Ensuring freedom to express views and organize for action.
- Giving a fair hearing to every stakeholders' grounds for entitlements, with no discrimination in favour of some with respect to others (discrimination may be based on ethnicity, gender, age, caste, economic power, religion, residence and so forth).
- Helping the stakeholders to participate in the negotiation process, for instance, by supporting them to organize, to develop a fair system of representation and to travel to meetings.
- Organizing discussion platforms where all the stakeholders can voice their ideas and concerns, selecting the least discriminatory places, times, languages, formats, etc.
- Supporting the negotiation of a fair share of management functions, rights, benefits and responsibilities.
- Ensuring effective and unbiased facilitation during negotiations.
- Supporting (via training and allocation of resources) the skills and capability of stakeholders to negotiate.
- Promoting a tight proportionality between the management entitlements and responsibilities and the benefits and costs assigned to each stakeholder.
- Keeping an open door for new stakeholders who may arrive on the scene.
- Supporting participatory democracy and multi-party agreements and organizations in all sorts of social decisions.
- Ensuring a fair measure of democratic experimentalism, allowing the adjustment of management plans, agreements, organizations and rules on the basis of learning-by-doing.
- Ensuring that the negotiated co-management plans, agreements and rules are enforced effectively.

Source: Borrini-Feyerabend *et al.* (2000, p. 33).

- Acknowledge that each group may not fully and unconditionally support proposed agreements. Encourage stakeholders to express their doubts about agreements. View 'consensus' as likely to mask differences in perspective and discount the input of disadvantaged groups.
- Assess the likelihood that agreements may need to be renegotiated in the future.
- Prepare disadvantaged groups to the possibility that the good will demonstrated among groups may not last.
- Stakeholders share information fully and openly.
- Assess the legitimacy of processes, decisions and agreements in terms of the role and implications for disadvantaged groups. Analyse the reasons for participation or non-participation of each group in negotiations, their objectives and strategies, how groups are represented, and the history of relationships underlying agreements.
- Open communication among the groups.

10.3. Establish a Co-management Body

A co-management body is established to conduct and oversee the planning process. The co-management body provides a discussion forum where different viewpoints are aired, options evaluated, conflicts managed, and plans and agreements negotiated. This may be a formal body or an informally organized or temporary body to facilitate the planning process. The co-management body may eventually become the co-management organization (see Section 10.12). Community representation in the management body is essential.

The co-management body is composed of representatives from the community, community organization(s), local government and the external agent/CO. Leaders of the co-management body are elected from among the representatives. The co-management body has the task of developing a mission statement; preparing the plan, including reviewing issues, problems and opportunities and defining goals, objectives and activities; holding community consultations on the plan; finalization, submission and adoption of the plan; and preparing a co-management agreement (Box 10.5)

The co-management body should hold frequent community consultations and meetings throughout the planning process to obtain input from the community on issues that develop in the process and to keep them informed, interested and active in the process. It is especially important for community members to participate in establishing the mission, goal and objectives for fisheries management. Decisions will need to be made about which meetings will be open to the public during the planning process and which will be closed, not for lack of transparency but to expedite decision-making.

Box 10.5. A Process of Co-management in Jamaica, Caribbean.

Since co-management is not a part of Caribbean custom or tradition, a proposed structure within which co-management can take place is a stakeholder council which can be called a Fisheries Management Council. The membership of the Council should be broad based with representatives of all interests in the fishery, such as cooperatives, organizations, commercial fishers, recreational fishers, Fisheries Department, Environment Ministry, Coast Guard, Marine Police, NGOs and marine research institutions. Each delegate must be selected by the group represented. There must be a process where the delegates communicate the decisions of the Council back to their organizations for discussion and ratification. The Council will look after resource monitoring, development of regulations, and enforcement. The first duties of the Council are to define the area to be managed and to draft the fisheries regulations to be implemented. The Council will need to seek funding for the programme.

Source: Espeut (nd).

10.4. Agreeing on Rules and Procedures for Negotiation

(The source for this section is: Borrini-Feyerabend *et al.*, 2000.)

All stakeholders should be sent an invitation and a proposed agenda for a first meeting of the co-management body. The first meeting should address procedural issues (Box 10.6). The first meeting should set a calm and productive atmosphere in order to help the stakeholders find out where they stand, establish working relationships, and start to own the process. Substantive issues should be addressed at subsequent meetings which may only be attended by co-management body members. The CO and other core group members can describe the work so far, including research activities, community organizing and funding support. Each stakeholder representative can introduce themselves and their organization and how they became the representative.

Box 10.6. An Example of a Set of Rules for the Negotiation Process.

This is only an *example* of a set of rules. Rules for negotiation processes are strongly dependent on the cultural setting.

- All main stakeholders should be present and participate through their formal representatives;
- Participation is voluntary;
- If more than X% of the stakeholders are not present for a meeting, the meeting will be adjourned;
- Language should always be respectful;
- Everyone agrees not to interrupt people who are speaking;
- Everyone agrees to talk only on the basis of personal experience and/or concrete, verifiable facts;
- Everyone agrees not to put forth the opinions of people who are not attending the meetings;
- Consensus is to be reached on all decisions and voting should be used only in exceptional cases;
- 'Observers' are welcome to attend all negotiation meetings.

Source: Borrini-Feyerabend *et al.* (2000, pp. 38–39).

The use of an external facilitator is recommended. The CO or a community member or leader could facilitate the meeting but an external facilitator brings an independent and objective perspective to the process. The facilitator can help:

- Define rules and procedures;
- Arrange logistics for meetings;
- Ensure that the process is transparent and representative;

- Ensure that everyone has a fair chance to participate;
- Ensure open communications;
- Provide information and options;
- By not stating his or her opinion and not deciding anything;
- Let everyone know when an agreement has been reached.

Those present at the meetings are representatives of the stakeholders in the co-management programme. It is the task of all representatives together to identify their role and responsibility in terms of substantive issues and decisions.

10.5. Meetings to Review the Situation and Develop Priority Issues

Several meetings may be needed to agree on the rules and procedures. Subsequent meetings should be aimed at establishing a basis of common interest and concerns among all the representatives. This can be facilitated by reviewing the community profile which provides detailed information on the ecological, economic, social and institutional/legal situation and trends, and problems, needs and opportunities.

The community profile should not define or limit the discussion. The facilitator can use participatory tools such as brainstorming, trend analysis, problem ranking, SWOT analysis, problem trees and webs, and site visits to work with the co-management body to better define the situation (Box 10.7). The facilitator helps the representatives develop a consensus on critical problems, threats, needs and opportunities.

The cumulative result of the analysis and assessment exercises is the identification of priority issues to be addressed in the co-management plan.

Box 10.7. Trend Analysis.

Trend analysis is used as part of an individual or group interview and consists of an in-depth discussion of specific issues or phenomena. The main purpose of trend analysis is to assess changes over time, and to raise the awareness of people about phenomena that accumulate slowly.

The participants in the exercise select the topic/subject to assess and identify one or more accurate indicators. The facilitator then asks the participants to say where they think they are now in relation to each indicator, where they were 5, 10, 20 years ago, where they think they will be in 5, 10, 20 years. Together with them, draw a graph of the trend for each indicator, or use some symbolic graph. Once the trends are clear, the facilitator asks the participants to discuss them.

Source: Borrini-Feyerabend *et al.* (2000, pp. 71–72).

10.6. Developing a Mission Statement

(The source for this section is: Margoluis and Salafsky, 1998, Chapter 2.)

Before developing a plan, it is important to have a clear understanding of the body's mission, that is, a vision for the future. A mission statement does not focus on specific details or tactics but rather on the desired outcomes and general strategy for getting there (Box 10.8). A mission statement describes:

- *Purpose:* What the body is seeking to accomplish.
- *Strategies:* The general activities or programmes the body chooses to undertake to pursue its purpose.
- *Values:* The beliefs which the members of the body have in common and try to put into practice while implementing the strategy.

A meeting or series of meetings are facilitated to develop the mission statement. While the co-management body may be the focus of this effort, the

---

Box 10.8. Mission Statement.

A mission statement describes the purpose of the organization in a few words. It provides an identity and unites the group's energy and enthusiasm. Powerfully written, it acts like a magnet pulling the group in the direction that it wants to follow (Almerigi, 2000). (Note: many Philippine groups would discuss a mission as the 'dream' of the organization so people can hold on to it even if there are many obstacles in their immediate and medium-term dealings.)

 To write a mission statement, the following questions should be asked:

- What do we do?
- Who do we do it for?
- How do we do it?
- Why do we do it?
- What are our values?

 A mission statement should:

- Be about who we are now, not what we want to be in the future;
- Be short, clear and usually less than 14 words;
- Stir up people's passion;
- Be connected to our deepest interests;
- Be a unique description of the organization; and
- Not be fuzzy. Avoid words that mean different things to different people such as excellent, best, etc.

 An example of a mission statement is:

 To improve the standard of living of fishers through education and sustainable fishing practices.

Source: Almerigi (2000).

community is involved through visioning workshops where they identify desired outcomes or vision for the fisheries (Box 10.9).

Box 10.9. Visioning Workshops to Support Sustainable Livelihoods in the Philippines.

Pamana Ka sa Pilipinas, Tambuyog Development Center (TDC) and the Network of Sustainable Livelihoods Catalysts (NSLC) are three groups in the Philippines working together to support community research and capacity building for sustainable livelihoods. It is usual for livelihoods and economic development to be treated as a 'component' of a CBCRM project; however, these groups would like to view livelihoods in relation to organizational development support and resource management activities. They are supporting visioning workshops where communities are able to articulate their personal 'dreams' and vision for their communities. The outputs of these processes vary. In the case of NSLC, the visioning exercises provide an understanding of community needs, interests and assets and are used in developing training modules on capacity building for sustainable livelihoods (SL). Tambuyog envisions these processes to feed into the barangay planning. In a PAMANA site in Bohol, the SL visioning exercise led to the development of a community plan that relates resource management activities and livelihoods activities.

The SL visioning processes are led by the external organizations but there are strong partnerships with local groups and selected community members who act as local researchers. In the case of Tambuyog, four local researchers from the fisher cooperative and women's organization co-facilitated the community orientation and the SL visioning exercises using PRA tools such as life histories on success, life cycle model analysis and asset pentagon mapping. NSLC calls the local researchers 'community scholars', following the example of Bolinao where fisherfolk leaders who were nominated by their organizations as community scholars subsequently learned the skills and attitudes of a community organizer with the intention of developing and strengthening local capacity. Called local community organizers (LCOs), these leaders learned through sustained sharing and learning-by-doing activities.

As an entry point for support work in the community, these visioning workshops focus on questions such as:

- Can you describe your community?
- Can you describe your life in the community?
- Can you describe the livelihoods found in your community?
- Can you describe your own livelihoods?
- How can life be improved in your community?
- What are the needs of the community?
- How can your life be improved?
- What do you need to improve your life in the community?
- What can the community do to improve?
- What can you do to improve your life in the community?

The central element of these visioning workshops is the discussion on 'livelihoods'. These Filipino groups emphasize that talking about livelihoods is not the same as implementing a livelihood project. The visioning workshops initiate a discussion about livelihoods and how they can be improved. It deepens a discussion about life itself and gives people a chance to understand their own livelihoods and make decisions about their future. Visioning workshops may lead to implementing a livelihood project, but that depends on the situation. In some cases, people may want to improve on their organization first before implementing a livelihood project. Holding visioning workshops for people to articulate their perspectives and dreams about their lives is not the same thing as implementing a livelihood project as an entry point. They are separate ideas. The latter is a narrow way of equating livelihoods with income-generating projects while the former takes on a more holistic view of livelihoods where people are fundamentally linked with resources.

Source: Programme reports and documents of Network of Sustainable Livelihoods Catalysts, Tambuyog and Pamana Ka, Quezon City, Philippines.

First, the representatives need to agree on the *purpose* – that is, what is hoped to be accomplished and what problems need to be solved to get there. For example, to improve fisheries management and increase people's standard of living. The purpose statement should indicate a change in status (to improve) and a stated problem or condition that needs to be changed (fisheries management, standard of living). The facilitator should work with the co-management body to state and agree on what the purpose should be.

Once a purpose is agreed upon, broad *strategies* need to be considered that will move the process forward – what will actually be done. For example, marine protected areas could be established at specific sites for marine conservation or there could be lobbying of the national fisheries agency for fisheries policy change. Although generally several strategies can be used to achieve a given purpose, only one or two strategies should be employed at the start. Each strategy should describe a specific set of actions that can be taken to achieve the stated purpose. The co-management body should brainstorm different strategies and rank each alternative based on: (i) what is needed to undertake the strategy; (ii) whether skills and expertise exist to do it; and (iii) whether or not it is enjoyable to do it. Each alternative can be ranked on a scale of 1 to 5 and the strategy(s) with the highest total is selected.

Once a purpose and strategy have been agreed upon, it is time to discuss *values* – the beliefs that will guide the work. Value statements should outline a belief that the body holds that will influence what it will do or what it will not do to achieve its purpose. Representatives should suggest and discuss values that are important to them. General agreement is reached on those values important to the co-management body. Unlike the purpose and strategy, there may be multiple values.

Once a purpose, strategy and values have been agreed upon, they should be written into a formal *mission statement*. The draft mission statement can be written by one or two people and circulated for review and comment. An example of a mission statement is:

> We, the people of Hung Thang village, wish to promote our health and well-being by finding ways to better manage our fisheries resources upon which our children and their children's livelihoods depend, now and in the future.

Once a mission statement has been agreed upon, it should be posted for the community to see and the representatives should discuss it with their organization's members. There should be broad public consensus and affirmation on the mission statement.

It should be noted that if substantial disagreement exists among the representatives at any point in developing the mission statement, the leaders will need to make a decision about the mission statement or the co-management body will need to be divided to make decisions and reach consensus.

Upon broad community agreement, there should be a ceremony where the co-management body representatives sign a written document that outlines the mission statement including what is to be achieved and how it will be accomplished. This should be done in writing to formally define the partnership and provide clarity. Such a ceremony helps raise the mission statement to a higher symbolic level, making it valid. With this common ground, it can help all stakeholders reconcile controversies and conflicts that may arise (Borrini-Feyerabend *et al.*, 2000).

10.7. Establishing the Management Unit

Fisheries management has traditionally focused on managing individual fish stocks. Recently, however, there is agreement that fisheries management requires the management of a larger coastal and marine area or ecosystem. This management unit needs to be clearly defined to include natural and human components, such as ecosystems, human uses and political boundaries (Berkes *et al.*, 2001). The management unit will be the legal area covered for co-management under the plan and agreement.

In the community entry and integration phase, area boundaries were identified (see Chapter 6, Section 6.2.5). These boundaries were mapped and include political, ecological, geological, fishing tenure, fishing gear area and management. More information was obtained in the research activities from the Resource and Ecological Assessment, Socio-economic Assessment and Legal and Institutional Assessment.

The co-management body will need to identify and select the management unit. A complex web of ecological, geological, social, cultural, economic and political factors influence the process of identifying and selecting the management unit. The management unit must be small enough to be easily managed but large enough to address ecological management issues. The larger the management unit and the more people involved, the more difficult management becomes. Starting small is often advisable.

At the initial establishment of the management unit the right scale of stakeholders and resources may not be identified. The spatial scale of the management unit can always be changed later as the co-management programme matures. The programme can be expanded to address new stakeholder or multi-municipal issues such as bay-wide management or upland or watershed issues.

In most cases, political boundaries will define the management unit. The smallest unit of government, municipality or district, in which the national government has already given some level of management responsibility over coastal waters, becomes the management unit. When possible, biological (fish migration, spawning area, coral reef) and social (use areas, tenure areas) considerations should be included in making a decision about the size of the management unit.

It will be necessary to secure legal protection for the management unit. It may be possible to approve ordinances at the local government level. In other cases, assistance from national government may be required.

10.8. Negotiating Co-management Plans and Agreements

(The source for this section is: Borrini-Feyerabend *et al*. 2000.)

As will be discussed below, the co-management plan is composed of a goal, objectives and activities. Before discussing each of these components, various methods and tools for stakeholders to agree on a course of action are presented.

For each component of the plan, the stakeholders have to identify what needs to be done to progress towards the desired future. The objectives have to be translated into activities and workplans that answer specific questions such as who needs to do what, by when, where, how, and with what financial and human resources. This is the moment when issues become concrete, numerous strategic options and choices are apparent to everyone, different points of view abound and conflicts surface (Box 10.10).

Box 10.10. Developing a Structure for Participatory Resource Management in Soufriere, St Lucia, Caribbean.

The Soufriere Marine Management Area (SMMA) in St Lucia evolved out of a process that began in 1992 to address growing environmental problems and conflicts related to the use of coastal and marine nearshore resources. Following a 2 year period of meetings and negotiations between stakeholders, an agreement based on a zoning plan was reached on the use and management of those resources. Conflicts persisted throughout 1996–1997 and it was felt that many of the difficulties experienced were the result of institutional inadequacies and a review was initiated. That process revealed a general consensus that the original institutional structure was not helping to achieve the objectives of effective management and ongoing consultation. Through a negotiation process among stakeholders, a new structure has emerged which builds their input into planning and decision-making. A new, formal co-management agreement has been signed by all partners. The original agreement was not a binding contract among the partners and the structure and operations of the Technical Advisory Committee were too cumbersome. The new agreement is guided by a clear mission, its structure is transparent and autonomous, and management enforcement is locally based.

Source: Krishnarayan *et al*. (2002).

Working groups can be formed for each activity, making sure that those most directly affected are represented in the relevant working group. It is also a good idea for each group to have its own facilitator or moderator, perhaps someone who takes on a neutral role learnt from the professional facilitator.

The groups need to come to terms with the various options open to them to achieve the same objective and select the one best suited to the conditions and needs of the given context. The tools already used to arrive at the mission statement (e.g. brainstorming, problem analysis, SWOT) can help again, but other techniques can also be used, such as:

- Discussing the hypotheses and basic assumptions relating to each option, i.e. why it is thought that a certain action will lead to a certain outcome, such as the probability that fish catches will increase if an area is closed to fishing.
- Comparing the expected impacts of various options, whether environmental, social or economic.
- Comparing the feasibility of various options, including questions of cost and timescales.
- Effective conflict mediation on the basis of an agreement that satisfies every party (a win–win solution) is likely to be more long lasting and more satisfactory than a win–lose situation.
- Facilitating the achievement of satisfactory compromises through the use of flexible instruments, such as spatial zoning and/or detailed conditions for the use of the resources at stake.
- Asking the stakeholders to devise (and estimate the feasibility) of incentives that would encourage them to agree on a given option.
- Calling for expert opinion on controversial issues.

If the small working group cannot arrive at a consensus on any of the options, it can present them all to the whole co-management body and ask for the advice of everyone. The body may compare the different options according to criteria such as cost, equity or ecological benefit, and may pursue the discussion, perhaps proposing additional compensation or incentives as one option. The aim is to achieve a consensus on the most appropriate course of action for each activity.

10.9. Co-management Plan – Goals, Objectives, Activities

(The source for this section is: Margoluis and Salafsky, 1998, Chapter 4.)

The co-management body has reviewed the situation and identified problems, threats, needs and opportunities. The challenge now is to design a management plan that will enable the community to address the situation. The management plan includes goals, objectives and activities (Boxes 10.11 and 10.12). Since there are usually limited funds and resources available to address needs and problems, it is important for the plan to be focused on specific strategies. In addition, goal and objectives need to be clearly stated.

Box 10.11. Co-management Plan Outline.

A co-management plan includes the following basic parts:

1. *Description of the area and resource.* This includes geography, demography, important coastal resources and their condition, socio-economic status of the people, institutions and laws, and other relevant information for management. Use graphs and tables to present baseline data.
2. *Maps of different scales.* Include a map of the entire area and detailed maps of the coastal area with resource locations and use patterns, existing management interventions and other data.
3. *Management issues/problems.* Priority issues must be clearly stated along with their contributing causes and factors. Trends in decline of resources can be used to illustrate issues of concern.
4. *Goals and objectives.* Management goals and objectives should be derived from the main issues.
5. *Activities.* Activities for each objective with assigned responsibilities should address each major issue at the heart of the plan. The functions and responsibilities assigned to each stakeholder. The activities can also be considered as the management measures and alternatives to be undertaken.
6. *Institutional and legal framework.* This section explains the framework that supports the plan, what institution is responsible, and how it is supported by the law. Procedures for negotiating ongoing decisions and managing eventual conflicts. Procedures for implementing and enforcing decisions.
7. *Timeline.* A schedule for implementation helps organize all responsible parties to implement the plan in a timely manner.
8. *Budget.* Funding needs and sources are identified for each activity.
9. *Monitoring and evaluation.* Monitoring and evaluation must be included as a set of activities to provide feedback on plan implementation and impact on the environment.

Source: DENR *et al.* (2001b, p. 54).

In developing the plan, a number of participatory tools and methods can be used including focus group discussions, problem trees and preference ranking. It is important for the whole community to be clear about the goal and objectives and about what can be achieved in order to focus strategies. The more participatory the process of setting goal and objectives, the greater the community acceptance and legitimacy of the co-management programme. Often the process of arriving at consensus regarding goal and objectives is an effective means of promoting an exchange of information and understanding among stakeholders. If a trained planner is not available, a plan can still be prepared based on the stakeholders' knowledge and participation.

As discussed in Berkes *et al.* (2001), fisheries management makes use of objectives in situations when a biological assessment, a Stock Assessment Driven management approach, is not feasible or affordable. Through a Management Objective Driven (MOD) approach, the focus of management is how to determine when the desired objective has been achieved; that is, how

---

Box 10.12. A Project Approach to Planning.

One approach to planning is to view management as a project. The value of approaching fisheries management planning as a project is the requirement that goal, purpose, objectives, activities, outputs and means of evaluation be clearly stated. It also puts the exercise in a specific time frame. *Logical Framework Analysis* (LFA) is a popular approach to developing, communicating and managing projects. Many donors require that projects submitted for funding are prepared in this format.

A full description of LFA would require more detail than can be provided here. Guides to the use of LFA in project planning have been developed by several agencies (Commission of the European Communities, 1993; USAID, 1994; IADB, 1997). The LFA consists of a series of processes:

- Stakeholder analysis;
- Problem analysis;
- Objective analysis;
- Analysis of alternatives;
- The logical framework matrix;
- Execution plan;
- Monitoring and evaluation plan;
- Project reports.

The logical framework matrix is peculiar to LFA. Its purpose is to summarize the project clearly and succinctly in a standard format. The rows of the matrix are referred to as the vertical logic. From the top down are:

- The goal that the project serves;
- The specific purpose of the project;
- The outputs that will be generated in order to achieve the purpose;
- The activities that will be carried out to generate the outputs.

The columns of the matrix are:

- The objectives;
- The indicators that the objectives have been achieved;
- The means of verifying the indicators;
- The assumptions upon which the achievement of the objectives is based.

The matrix also summarizes the resources that would be needed to produce the outputs.

Source: Berkes *et al.* (2001, pp. 46–48).

---

to measure successful management. This approach requires the knowledge and information to identify variables that relate to the objectives, followed by the setting of target points on those variables. This MOD approach ensures that the management system focuses on the acquisition and analysis of data that relate to the objectives and control system. (For more on fishery management objectives see Berkes *et al.* (2001, Section 3.6, pp. 58–66).)

A *goal* is a general statement of the desired state towards which the stakeholders are working. A good goal meets the following criteria:

- Visionary: inspirational in outlining the desired state towards which the stakeholders are working.
- Relatively general: broadly defined to encompass all activities.
- Brief: simple and succinct so that all stakeholders can remember it.
- Measurable: defined so that changes can be accurately assessed.

Although in some instances there may be multiple goals, it is usually easier to have only one goal. An example of a goal is:

> To ensure the availability of fish for our grandchildren and our grandchildren's grandchildren

Review the goal with the management body and put it in a place where everyone can see it on a regular basis and gain inspiration from it.

Objectives are specific statements detailing the desired accomplishments or outcomes to be achieved (Box 10.13). Realization of objectives should lead to the fulfilment of the goal. A good objective meets the following criteria:

- Impact oriented: represents desired changes in critical factors that affect the goal.
- Measurable: definable in relation to some standard scale (numbers, percentages, fractions or all/nothing states).
- Time-limited: achievable within a specific period of time.

Box 10.13. Objectives.

Some examples of objectives are:

Objective 1: Restrict location of fish harvesting

During the second year of the project there are no incidences of community members fishing in the sanctuary areas.

Project Assumptions: If community members are involved in establishing the sanctuary and are responsible for its success, then there will be no violations in the future.

Underlying Assumptions: (i) Setting up and maintaining the sanctuary areas on the coral reef will enable populations of marine organisms to maintain themselves and recover. (ii) Community members will respect and maintain the sanctuary areas.

Objective 2: Restrict timing of fish harvesting

By the end of the second year of the project, community leaders do not hear any substantiated reports of community members fishing for restricted species during their critical breeding periods.

Project Assumptions: The community leaders assume that if community members are made aware of the need to suspend fishing during critical breeding periods, they will respect the closed periods.

Underlying Assumptions: (i) Setting up and maintaining closed periods for fish harvesting during critical breeding periods will enable populations of fish to maintain themselves. (ii) Sufficient information is available on fish species' life cycles to be able to determine appropriate breeding periods.

- Specific: clearly defined so that all people involved have the same understanding of what the terms in the objectives mean.
- Practical: achievable and appropriate with the context of the community.

For priority issues identified earlier, write down one or more draft objectives. Each priority issue can have more than one objective associated with it, depending upon what is to be accomplished. There are trade-offs between having too many objectives and having too few. If objectives are getting too long and complex, divide them. A general rule is to have between one and three objectives for each priority issue. Once an objective has been written, it is useful to write down some notes outlining why and how the objective was developed. Also, write down all the assumptions that describe why the objective was developed.

Activities are specific actions or tasks undertaken to reach the objectives (Boxes 10.14 and 10.15). A good activity meets the following criteria:

- Linked: directly related to achieving a specific objective.
- Focused: outlines specific tasks that need to be carried out.
- Feasible: accomplishable in light of resources and constraints.
- Appropriate: acceptable to and fitting with community cultural, social and biological norms.

For each objective, write down a list of the major activities that will help

---

Box 10.14. Activities and Plan Implementation.

The development of a co-management plan is the starting point for improving the conditions of the fisheries, the community and individual households. The co-management plan serves as the basis for implementation, that is, the set of activities by which the plan and projects are carried out. The identified objectives may address both fisheries and non-fisheries elements.

Co-management can serve as a mechanism not only for fisheries management but also for community and household economic and social development. Many of the solutions to fisheries problems will lie outside the fisheries sector. As such, the activities identified in the co-management plan may include:

- Fisheries and coastal resource management;
- Community and economic development;
- Alternative and supplemental livelihoods;
- Continued social empowerment and capability building;
- Advocacy and networking;
- Institution building.

Examples of the activities identified may include fisheries resource management, such as marine protected areas and improved enforcement. They may also include the development of alternative or supplemental livelihoods for fishers and their families and community development activities such as building a water well or improved education for children.

Box 10.15. Objective and Activities.

An example of an objective and activities:

Objective 1: Restrict location of fish harvesting

Activity 1. Hold meetings with community members

Hold meetings with community members to discuss declaring part of the fishing grounds to be sanctuary areas.

- Why do this activity? No major decisions are taken in the community without full participation of all community members.
- How will the activity be carried out? Co-management organization leaders will hold meetings with all community members to explain and get agreement on sanctuary.
- Who is responsible for the activity? Mr Leonard George and Mr Randolph McKinney, representatives of the Point Fishers Association.
- When will the activity take place? Second month of the project.
- Where will the activity take place? Primary school.
- Underlying assumptions? None.
- Prerequisites? None.

Activity 2. Designate sanctuary areas

Convene meeting of co-management organization to designate, map and mark sanctuary areas based on traditional fishing grounds.

- Why do this activity? In past community meetings, some residents have strongly supported the idea of setting aside sanctuary areas where fish can safely grow and reproduce. At present, however, there are no areas in the region where there are restrictions on fishing.
- How will the activity be carried out? The co-management organization will convene a meeting and solicit suggestions (based on the results of the community meetings) as to which areas should be designated sanctuaries and how they should be marked.
- Who is responsible for the activity? Mrs Harriet Mahon, representative of Bay Fish Traders Association.
- When will the activity take place? Fourth month of the project.
- Where will the activity take place? Primary school.
- Underlying assumptions? Community members will be able to agree on sanctuary sites.
- Prerequisites? Discuss with community before meeting.

it to be achieved. Try to arrange the activities in the sequence in which they will need to occur to achieve the task. Improved fisheries may require a long string of activities over an extended period of time before the desired objective is achieved. For each activity include:

- Who will do the activity?
- How will the activity be carried out?
- Who is responsible for the activity?
- When will the activity take place?

- Where will the activity take place?
- Underlying assumptions;
- Prerequisites.

Each activity and its associated task must be assigned to a specific person or persons to be completed by a specific date in a specific location. Each task must also have a budget to pay for it.

Once the goal, objectives and activities are identified, they are combined into a co-management plan. The simultaneous use of multiple types of activities is often required for effective management since no single strategy is likely to completely address a problem. To determine the appropriateness of any one or a combination of activities for any particular management problem requires more detailed analysis of the actions needed to fully implement the activity.

Tasks should be organized in a *timeline* which shows at what point in time each task should be undertaken and who is responsible for each task. Timelines can be prepared for the programme as a whole or for individual people within the programme.

A *budget* should also be created. The budget can be organized by objective, with each activity having an associated cost for staff salaries, equipment, supplies, travel and so forth. The subtotal cost for each objective as well as the total programme cost should be calculated. Income and costs should be compared to ensure that there are enough funds to undertake the programme.

In summary, the management plan summarizes all the activities and indicates (DENR *et al.*, 2001b):

- What is the goal and objective to be achieved?
- What is the specific activity for achieving the goal?
- What is the expected output (what will be produced or changed)?
- What is the timeframe (over what period will this strategy be implemented)?
- Who are the participants (whose cooperation is needed to effectively carry out this action)?
- Who is the lead person or organization (who is responsible for implementation)?
- What are the funding needs and sources?

10.10. Evaluation and Monitoring Plan

(The source for this section is: Margoluis and Salafsky, 1998, Chapter 4.)

10.10.1. Evaluation

Managing fisheries resources is a continuous, iterative, adaptive and participatory process comprised of a set of related tasks that must be carried out to achieve a desired set of objectives. Plans must be monitored if they are to be kept on track, and evaluated if there is to be learning from successes and failures. The planning cycle includes the process of assessment, monitoring and evaluation. Effective plan assessment and evaluation involves several steps: (i) preliminary appraisal; (ii) baseline assessment; (iii) monitoring; and (iv) evaluation. Information for each of these steps is essential to maximize chances that the plan will be effective for the adaptive management process and to acquire lessons learned.

Evaluation consists of reviewing results of actions taken and assessing whether these actions have produced the desired outcomes – this helps to adapt and improve by learning. Evaluation is a routine part of the adaptive management process and is something that most fisheries managers already do where the link between actions and outcomes can be simply observed. However, the links between actions and outcomes is often not so obvious. Faced with the daily demands of their jobs, many fisheries managers are not able to monitor systematically and evaluate the results of their efforts. In the absence of such evaluations, resources can be wasted on activities that do not achieve the objectives. Evaluation can be:

- *Formative or ongoing evaluation.* A continuous, ongoing monitoring and evaluation process during the implementation process where indicators of performance are obtained and systematically compared with objectives. The results are used for taking appropriate actions to make improvements.
- *Summative or post-evaluation.* Undertaken after the project's implementation where the focus is on deeper analysis of results and outcomes and for determining the level of achievement of objectives and the impact of the project. The results are used for future planning.

Monitoring and evaluation are processes which assist in answering the questions: Are the activities working or not? And what actions are needed to make them work? Monitoring answers the question: How are we doing? Evaluation answers the question: How did we do? If the plan has measurable objectives and indicators to evaluate the plan, ongoing monitoring can provide information required to evaluate effectiveness and performance of the plan. (For more information see Berkes *et al.* (2001, Chapter 5, pp.101–128).)

There are several approaches that may be used to evaluate co-management plans at the local level:

- Performance evaluations are designed largely to determine the quality of implementation of specific activities and the degree to which the goal and objectives are achieved. There is a focus on accountability, meeting budget commitments, quality control and terms of reference.
- Process evaluation examines the means by which goal and objectives are achieved. The evaluation is used to assess such means as the clarity of goal statements and legislative mandates, measures of rationality of organizational structures and the process of information flow, adequacy of yearly budget allocations, number of permits issued/denied and availability of trained staff.
- Management capacity assessments are designed to evaluate the adequacy of

plan implementation, policy framework and supporting institutional structures against a set of structures. The focus is on evaluating governance processes, management structures, policy tools, management options and strategies, regulatory mechanisms and policy enforcement, and human and institutional capacity.

- Outcome evaluations assess the socio-economic and environmental impacts of a co-management plan. The focus is on measured impacts on people and the environment and generally requires rigorous scientific methods to distinguish between outcome of interventions and other variables that may contribute to a measured outcome. Environmental or socio-economic conditions measure such things as the extent of protected habitat or the number of jobs created.

10.10.2. Monitoring plan

Adequate monitoring of co-management plans allows fine-tuning of project strategies and activities to more effectively respond to both environmental and human impacts. The preliminary appraisal and baseline assessment are done early on during the community entry and integration and research activities. Information from these activities serves both the monitoring and evaluation steps. Monitoring makes it possible to learn if fisheries management measures and alternatives are working or not.

The monitoring plan is the outline of the steps to undertake to ensure that the programme is on track. If monitoring is not undertaken, it will not be possible to know whether the goal and objectives are being achieved or what needs to be done to improve the programme (Box 10.16). In a general sense, monitoring provides a way of establishing the success of management measures. There are two primary reasons to monitor a programme. The first is to convince other people that you are doing what you said you would do. This type of monitoring is typically done to satisfy donor requirements or to help conduct a performance evaluation. The second reason is to learn whether the actions taken are working or not working so that corrective action can be taken if needed. With the second reason, monitoring is undertaken to determine whether the programme is being effective and to learn how to improve.

Box 10.16. Monitoring Strategy.

Once information needs have been identified, a monitoring strategy is designed to meet those information needs. For example, for:

Objective 1: Restrict location of fish harvesting. During the second year of the project, there are no incidences of community members fishing in the sanctuary areas;

The monitoring strategy would be:

Monitoring Strategy: Compare the number of sanctuary fishing violations over time.

Monitoring can be used to track process (governance aspects of co-management such as how planned activities are proceeding and level of participation) and results (outcome or impacts of the processes on behaviour change and socio-economic and biophysical conditions). Monitoring can show the community that their efforts are having positive effects on the resource they are managing. This will sustain and encourage participation by community members.

Monitoring data can be used for:

- Improving management decisions;
- Measuring project progress;
- Assessing effectiveness or impact of an activity;
- Determining lessons learned that can be used in future activities.

Baseline data, collected from the REA, SEA and LIA (community profile), at the start of implementation provide a basis for comparison. Baseline data are essential to the interpretation of monitoring data because they provide a 'before' situation from which to measure progress. Baseline or control sites are used to eliminate the normal range of variation encountered due to change. The variables to measure and the desired level of information to be collected must be balanced against cost. Sampling procedures will differ between environmental and socio-economic data.

It is important to note that the timescale of many fisheries management projects is very long, that is, changes may not be apparent for many years. Thus, monitoring should have both short-term (1–3 year) and long-term (10–20 year) time frames.

Monitoring should, wherever possible, be done by the community, not by outsiders. This is called Participatory Monitoring and Evaluation (PM&E) (Vernooy, 1999). PM&E should be seen as a learning process to build skills and knowledge. In general, monitoring activities are more effective as a means of feedback and encouragement if they are designed, implemented and interpreted by the community. The community may need assistance to develop a monitoring plan, identify indicators and interpret the results. A participatory monitoring programme may take longer to establish but creates ownership, skills, confidence and credibility in the community.

The first step in developing a monitoring plan is to determine the audience for the information generated by monitoring. The audience can be internal (members of the co-management organization) or external (donors, policy-makers). The priority need for information is to monitor the project's progress in achieving the goal and objectives. Additional information needs may include problems or threats that are not being directly addressed by the project but have an impact on meeting the objectives. For example, the international demand and trade for a certain fish species may be driving illegal and over-fishing in the area. While it is beyond the ability of the project to address this threat, information may be monitored to see if it does become necessary at some time in the future to design activities to deal with it.

Once what needs to be known has been determined and the general monitoring strategy has been identified, the next step is to develop specific *indicators* for each information need that will be followed throughout the life of the project. An indictor is a unit of information measured over time that documents changes in a specific factor (Box 10.17). A given goal or objective can have multiple indicators. The best indicators are those that are linked to the goal and objectives so that assessing them helps to measure reduction in the problem or threat. A good indicator meets the following criteria:

- Measurable: able to be recorded and analysed in quantitative or qualitative terms.
- Precise: defined the same way by all people.
- Consistent: not changing over time so that it always measures the same thing.
- Sensitive: changing proportionately in response to actual changes in the item being measured.

Box 10.17. Indicators.

For the example in Box 10.16, an indicator is:

Indicator 1: Number of reports of violations.

In some cases, the selection of the indicator is directly related to the goal or objective, in other cases it may require a bit more thought to develop the indicator.

Once the indicators have been selected, the next step is to select methods that will be used to collect data to measure the indicator. There is usually a wide range of methods which can be used to assess an indicator. As such, it may be useful to select the specific method based on certain criteria:

- Accuracy and reliability: how much error exists in data collected by using the method? To what degree will results be repeatable?
- Cost-effectiveness: what does the method require in terms of resource investment? Are there cheaper ways to get the same data?
- Feasibility: do people exist in the community who can use the method?
- Appropriateness: does the method make sense in the context of the project? Is it culturally suitable?

For each method, tasks that need to be completed to collect the data need to be identified. These include:

- Indicator: number of reports of violations;
- How: key informants (interviews with police who receive reports of violations);
- When: monthly;
- Who: one co-management organization member;
- Where: co-management organization monthly meeting make a report.

10.11. The Co-management Agreement

(The source for this section is: Borrini-Feyerabend *et al.*, 2000.)

The co-management plan developed by the co-management body needs to be binding. This is done by means of a formal agreement by the group of involved stakeholders and other external institutions (e.g. national government) as necessary (Box 10.18). The co-management plan should specify a share of functions, roles, benefits and responsibilities and be signed by each stakeholder group/partner involved. The more stakeholder groups and the more finances involved, the more advisable it is for the agreements to have a legal basis (e.g. as contracts). The signatories should be those stakeholders who are directly assigned a role and responsibility in the agreement.

All agreements should specify the activities to be undertaken, by whom and how, as well as the anticipated results and impacts to be monitored. They should also specify when those involved will meet again to assess whether the action has been effective and/or needs to be adjusted (evaluation).

Once the points of agreement are determined, all representatives must go to their constituencies for ratification. Copies of the agreement – written simply and in the local language – should be disseminated to the stakeholders and the public at large. It is important to keep the public informed on everything that happens in the negotiation meetings, and explain why agreements are reached

---

Box 10.18. Elements of a Co-management Agreement.

The co-management agreement may include the following elements:

- Identification of the co-management stakeholders/partners;
- Description of the project;
- Duration of the agreement;
- Description and identification of the management committee;
- Obligations of the government partner(s);
- Obligations of the non-government partner(s);
- Roles and responsibilities of each stakeholder/partner;
- Auditing and monitoring arrangements;
- Publication of particulars of the agreement;
- Termination provisions;
- Indemnification for third party liabilities;
- Indemnification of each other;
- Notices from the stakeholders/partners;
- Dispute resolution provisions;
- Applicable laws;
- Funding and resource provisions;
- Amendment provisions;
- Meeting schedule.

Source: National Roundtable on the Environment and the Economy (1998, p. 48).

> **Box 10.19.** Co-management Agreement in the Olifants River Harder Fishery, South Africa.
>
> Workshops were held over a 3-year period between the provincial Department of Cape Nature Conservation (CNC) and the fishing committee (which represented the broader fishing community) to identify the capabilities of the respective partners to undertake particular functions and on clarifying decision-making powers. Once both groups agreed on the division of management duties and responsibilities, the next step in the process was to finalize these arrangements in a formal 'partnership agreement'. The project team agreed to assist in preparing a draft agreement and facilitating the process further, if required.
>
> The draft partnership agreement was discussed at several workshops with members of the fishing committee and CNC. The intention was to involve all fishers in this partnership agreement formulation process and arrive at a consensus document that would be legally binding. Various legal mechanisms were considered. One such mechanism was the establishment of an Environmental Management Cooperation Agreement under the National Environmental Management Act 107 of 1988. In addition to clarifying powers and functions of respective partners, other important management considerations such as conflict resolution procedures, as well as how clauses would be amended, were addressed.
>
> Source: Sowman (2003).

on certain options. The social communication system set up earlier will be useful for this purpose (Box 10.19)

10.12. The Co-management Organization

As co-management is an adaptive process, the negotiation and implementation of the co-management plan and agreement is never finished. A relatively stable organization is needed to be in charge of the overall co-management programme (Borrini-Feyerabend *et al.*, 2000).

A co-management organization is established with the responsibility of managing the fisheries and to sustain the co-management programme, including the plan and agreement through time (Box 10.20). It has a mix of decision-making, advisory, operational and coordinating responsibilities. This is a permanent body. The functions of the co-management organization include (Heinen, 2003):

- Conflict management – to discuss and resolve conflicts among stakeholders;
- Policy-making – to prevent conflicts by translating the plans and agreements made into rules with appropriate penalties;
- Implementation – to ensure that management measures are followed by allocating funds and assigning people to different activities;
- Monitoring – to keep track of the effects and impact of the management measures;

Box 10.20. Co-management Organizations.

There may be different types and functions of co-management organizations depending upon the situation:

- Executive bodies (responsible for implementing plans and agreements on the basis of decisions produced by others, e.g. an association of local businesses responsible for executing a project negotiated between the director of a protected area and the bordering communities).
- Decision-making bodies (fully responsible for the management of a given territory, area or set of resource, e.g. the Co-management Board in charge of a defined area or the committee in charge of a Community Investment Fund).
- Advisory bodies (responsible for advising decision makers, e.g. a Coastal Council, directly linked with the regional authorities charged with the natural resource management mandate).
- Mixed bodies (for instance holding partial management responsibility and partial advisory responsibility, such as an Advisory/Management Committee responsible for advising a Marine Park Director on the decisions to be taken in park management but fully in charge of decisions and activities pertaining to the areas at its periphery).

The stakeholders may decide to set up several co-management organizations, for instance an advisory body and a management body.

Source: Borrini-Feyerabend *et al.* (2000, p. 53).

- Revising co-management plan and agreements – to sustain and update plan and agreements;
- Financing and fund raising;
- Information and data collection and analysis;
- Education;
- Research.

The co-management body established earlier to conduct and oversee the planning process may serve as the foundation for the co-management organization or a totally new organization may be established.

The membership may be representatives of all stakeholder groups, representatives of only a few of them, or professionals who do not represent any of the stakeholders. The co-management organization, composed of stakeholder representatives, may serve to make all day-to-day decisions or the stakeholder representatives may only serve to make policy decisions and leave the day-to-day decisions to hired professionals. Composition of the co-management organization is crucial. It is important to know who is being represented and what the balance of power is among the different stakeholders (Borrini-Feyerabend *et al.*, 2000). The most important requirement is that the stakeholders trust the people composing the co-management organization (Box 10.21).

The size of the co-management organization can vary depending on the stage of maturity of the co-management process. A small-sized co-

---

Box 10.21. Structure of Community-based Fisheries Management in Cambodia.

The Community Fisheries Committee is a part of the Community Development Committee of the village. The members of the Community Fisheries are elected from the local community and play a major role in managing the sustainable use of fisheries resources. The Community Fisheries reports to the Provincial Fishery Office and the village local authority. It is composed of:

- The Chairman – responsible for organizing the regulations and planning and implementation within the whole community. Arranges meetings, networking and communication with technical agencies, provincial authority and international organizations. Monitors the management of the fishery. Disseminates information about fisheries management to the community members and resolves conflicts.
- The Deputy of Community Fisheries Committee – takes on the work of the Chairman in his or her absence.
- Team Leaders – guide team members in implementing their work, gives information to members, monitors activities of the team, and develops ideas for action plans.
- Community Fishery Members – implement tasks of the team leader and Committee. Participate in voting to select committees and establish fishery regulations, and provide ideas to improve the fishery.

Source: Khai Syrado and Thai Kimseng (2002).

---

management organization can manage a well-established system. However, a new co-management process has to involve the participation of as many stakeholders as possible. As a result, a co-management organization in its early stage may initially be composed of 10–20 people/representatives. Once the rules and systems are in operation, the co-management organization can be reduced if efficiency so requires. If appropriate and depending on finances, representatives may be compensated (Heinen, 2003).

In addition to the size of the co-management organization, decisions will need to be made concerning whether or not it will be legally recognized or informal. In addition, internal rules such as election of officers, length of membership, chairmanship, meeting schedule, training of members, decision-making by consensus, recourse of mediation or arbitration, reporting rules and other management issues will need to be decided upon. In situations where the stakeholders share the same norms and values, consensus building may be an appropriate management style. In situations where there is a lack of these shared values and behaviour, stronger leadership and power may be a more appropriate management style.

As already mentioned, it is essential that the co-management organization earns and maintains the trust of the stakeholders. To be effective, there must be open communication, transparency and accountability in all the decisions taken by the co-management organization.

Questions of finances are always crucial. How will the co-management organization sustain itself in the future? Does it have its own assets or are there income-generating activities? While the co-management process and organization may be financed initially from outside donor funding, this probably should not be considered as a secure long-term source of financing. Some type of self-financing mechanism will need to be established. This may include user fees, membership fees, trust fund, government line-item budget or other source.

10.13. Revenue Generation and Financing

Co-management requires substantial financial resources to support the programme. Funds need to be available to support various operations and facilities related to planning, implementation, coordination, monitoring and enforcement, among others. Funding, especially sufficient, timely and sustained funding is critical to the sustainability of the co-management programme. In the early stages of implementation funding may have been obtained from a donor organization or a large development project. This source of funding may or may not continue in the long run. Programmes often fail when this outside funding stops. Funds also need to be made available on a timely basis to maintain cash flow for such things as staff salary and activities. The co-management programme must be supported and accepted by the community so that stakeholders will be confident enough to invest their own time and funds.

The co-management programme must be designed from the start with thoughts and plans for financing. Too much dependency on external sources will impact upon sustainability. As such, the co-management organization will need to consider how to generate revenue to finance the activities. Several sources of financing may be required. Financing mechanisms should be evaluated as part of a business plan that includes a sustainable financing strategy. The business plan should be based on an evaluation of the costs of operations for activities in the management plan. A range of potential alternatives can then be identified as potential financing sources for co-management. The choice of which financing mechanism(s) to utilize in a particular case should be based on analysing several feasibility factors (Spergel and Moye, 2004):

- Financial (funding needed, revenue generation, revenue flow, year-to-year needs);
- Legal (legal support for financing mechanism, new legislation needed);
- Administrative (level of difficulty to collect and enforce, complication and cost; potential for corruption, staff requirements);
- Social (who will pay, willingness to pay, equity, impacts);
- Political (government support, monitored by external sources);
- Environmental (impact).

Depending upon the situation, and the relationship with local government, several sources may be available (Spergel and Moye, 2004):

- Government revenue allocations;

- ○ Direct allocations from government budget;
- ○ Government bonds and taxes earmarked for conservation;
- ○ Lottery revenues;
- ○ Wildlife stamps;
- ○ Debt relief.

- Grant and donations

  - ○ Bilateral and multilateral donors;
  - ○ Foundations;
  - ○ Non-government organizations;
  - ○ Private sector;
  - ○ Trust funds.

- Tourism revenues

  - ○ Fees (entry, diving, yachting, fishing);
  - ○ Tourism-related operations of management authorities;
  - ○ Hotel taxes;
  - ○ Visitor fees and taxes;
  - ○ Voluntary contributions by tourists and tourism operators;
  - ○ Cost recovery mechanisms.

- Real estate and development rights

  - ○ Purchases or donations of land and/or underwater property;
  - ○ Conservation easements;
  - ○ Real estate tax surcharges for conservation;
  - ○ Tradable development rights and wetland banking;
  - ○ Conservation concessions.

- Fishing industry revenues

  - ○ Fish catch and services levies/cost recovery mechanisms;
  - ○ Eco-labelling and product certification;
  - ○ Fishing access payments;
  - ○ Recreational fishing licence fees and excise taxes;
  - ○ Aquaculture licence fees and taxes;
  - ○ Fines for illegal fishing.

- Energy and mining revenues

  - ○ Oil spill fines and funds;
  - ○ Royalties and fees for offshore mining and oil and gas;
  - ○ Right-of-way fees for oil and gas pipelines and telecommunications infrastructure;
  - ○ Hydroelectric power revenues;
  - ○ Voluntary contributions by energy companies.

- For-profit investments linked to marine conservation

  ○ Private sector investments promoting conservation;
  ○ Biodiversity prospecting.

- Other sources

  ○ Loans;
  ○ Income derived from local enterprises such as handicrafts, aquatic products, visitor gifts (t-shirts).

10.14. Legal and Policy Support

The success of the plan and agreements depends not only on the research, organizing and planning activities, but also on the legislation which provides the legal basis for the plan and agreements. There is a need to review and confirm the legal basis for all plans, agreements and proposed activities at local, national and international levels. Without such a legislative framework, there is no basis within which policies can be formulated and actions taken. For the fisheries to be managed with some semblance of order, its basic principles should be expressed in the form of law and policy. Key national laws and policies, which should have been identified in the Legal and Institutional Assessment, are again analysed to determine support for the activity. Where devolution to local governments has occurred, the mandates and authorities of the local government are also analysed.

In some cases, there is an absence of a legal and institutional framework for co-management and certain activities. There may be a need for new national legislation and/or for ordinances to be adopted at the local government level to support an activity. These should be prepared and reviewed with officials and legal council. In addition, the legal framework for such issues as customary/traditional management arrangements, resource access/tenure rights and indigenous people's rights need to be reviewed and clarified. It is important to determine the constitutional and legal aspects of, for example, devolving rights to organize and manage resources to community organizations.

An ordinance to support a specific activity, such as a marine protected area, may start as a resolution in a committee of the co-management organization (Casia, 2000). The resolution is a recommendation or an expression of intent. It is not recognized as law until it is approved, in the form of an ordinance, at the local government level. The resolution receives a public hearing so that community members can provide inputs on the issues to be addressed by the proposed law. Most ordinances will include the following basic elements (Casia, 2000):

- Declaration of policy;
- Definition of terms;
- Prohibitions;

- Penalties;
- Exemptions.

10.15. Publicizing

The end of the negotiation process is usually a meeting in which the results are made known to the community. In addition, authorities from outside the community may be in attendance. At the meeting, the agreements are presented for all to see. This is an opportunity to celebrate the work that has been accomplished. All those involved are acknowledged. During this celebration, the organizations established and the plans and agreements made are reconfirmed by all the stakeholders.

11 Conflict Management

L. Bunce.

(A primary reference for this section is: Rijsberman, F. (ed.), 1999.)

A classic book on conflict management (Moore, 1986) opens with:

> All societies, communities, organizations, and interpersonal relationships experience conflict at one time or another in the process of day-to-day interaction. Conflict is not necessarily bad, abnormal, or dysfunctional; it is a fact of life.

(p. ix)

Conflicts over fisheries and marine resources have many dimensions including, but not limited to, power, technology, political, gender, age and ethnicity. Conflicts can take place at a variety of levels, from within the household to the community, regional, societal and global scales. The intensity of conflict may vary from confusion and frustration over the directions fisheries management is taking to violent clashes between groups over resource ownership rights and responsibilities. Conflict may result from power differences between individuals or groups or through actions that threaten livelihoods (Buckles and Rusnak, 1999).

Buckles and Rusnak (1999) report that the use of natural resources is susceptible to conflict for a number of reasons:

- Natural resources are embedded in an environment or interconnected space where actions by one individual or group may generate effects far off-site.
- Natural resources are embedded in a shared social space where complex and unequal relations are established among a wide range of social actors – fishers, fish traders, boat owners, government agencies, etc. Those actors with the greatest access to power are also best able to control and influence natural resource decisions in their favour.
- Natural resources are subject to increasing scarcity due to rapid environmental change, increasing demand and their unequal distribution.
- Natural resources are used by people in ways that are defined symbolically. Aquatic species and coral reefs are not just material resources people compete over, but are also part of a particular way of life, an ethnic identity and a set of gender and age roles. These symbolic dimensions of natural resources lend themselves to ideological, social and political struggles that have enormous practical significance for their management and the process of conflict management.

Buckles and Rusnak (1999) further state that because of these dimensions of natural resource management, specific natural resource conflicts usually have multiple causes – some proximate, others underlying or contributing. A pluralistic approach that recognizes the multiple perspectives of stakeholders and the simultaneous effects of diverse causes in natural resource conflicts is needed to understand the initial situation and identify strategies for promoting change.

Conflict management is about helping people in conflict develop an effective process for dealing with their differences (Box 11.1). The problem lies in how conflict is managed. The generally accepted approach to conflict management recognizes that the parties in a dispute have different and frequently opposing views about the proper solution to a problem, but acknowledges that each group's views, from the group's perspective, may be both rational and legitimate. Thus, the goal of people working in conflict management is not to avoid conflict, but to develop the skills that can help people express their differences and solve their problems in a collaborative way.

The general reason for attempting a voluntary, collaborative approach is that it is often very costly, if it is at all possible, to resolve disputes through the courts and such legal proceedings can lead to delays, increased conflict, and outcomes that are unsatisfactory to one or all parties involved. The legal system is also not very good at finding creative solutions to conflict. Thus,

Box 11.1. Conflict and Community Law Enforcement in Bohol Province, Philippines.

Conflicts of various forms and ways are real problems in resource management. The experience in Bohol province, Philippines, shows how multiple stakeholder participation could help address conflict situations, and in particular, how direct action could be a potential model for community law enforcement initiatives.

In Bohol, an Environmental Summit was held in 1997 that led to the signing of a Bohol Covenant for Sustainable Development and, subsequently, the adoption of an Environmental Code for the province in 1998 and the establishment of the Bohol Environment Management Office. This office is tasked to facilitate, integrate, coach and harmonize activities on environmental protection, conservation and management. It also mandated the creation of a multi-sectoral task force focused on coastal resource management issues. This task force called for a Coastal Law Enforcement Summit in May and June 2000 that led to the creation of the Coastal Law Enforcement Council (CLEC). A memorandum of agreement was made between League of Municipalities of Bohol, national government agencies and ELAC to formalize the organization of the CLEC per congressional district.

The CLEC is a multistakeholder body from the FARMC, police, fisheries and environmental offices of the government, municipal councils including the mayors and vice-mayors, coast guard and the Environment Management Office of Bohol. It has an advisory group that includes a congressman, the Director of the Department of Interior and Government, representatives from the judiciary, NGOs and CRM projects in the province. The CLEC is mandated to identify the base of operations for the district, organize a composite coastal law enforcement team including advisers, produce a district-wide coastal law enforcement communications and operations plan, procure budgetary allocations and logistics, prepare a training and capability building programme, standardize policies and coordinate with coastal local governments.

The experience of CLEC in Bohol demonstrates that it is possible to manage conflicts arising from resource use if communities are involved and if there is strong support from multistakeholders. The CLEC as an enforcement body is challenged by inadequacy of financial and technical resources, but their work continues by encouraging closer coordination among stakeholders and sustained environmental education work.

Source: Piquero (2004).

alternative approaches to the courts are sought that may be faster, cheaper, and more effective and acceptable. In other situations, where legal action may not be an option, the alternative of not finding a solution at all may also be costly, if only in terms of opportunities lost.

The emphasis on the word 'voluntary', or mutually agreed upon, refers to the fact that conflict management approaches will only work if all parties to the conflict are convinced that they will be, or at least may be, better off by participating than they would be otherwise. This implies that as long as one of the parties feels that it can force its own solution, or could obtain a total victory at acceptable costs through the courts, or would actually benefit from no action, then conflict management approaches will not work. In conflict management terminology, this concept is referred to as BATNA, the best alternative to a negotiated agreement. As long as any of the parties in a conflict perceives its BATNA to be superior to participation in a conflict management approach, it may refuse to participate in such a process. One party's decision not to participate does not necessarily mean that conflict management is impossible. It depends on whether that party controls resources that are essential to dealing with the conflict or has effective 'veto power' over the agreement that the other parties might reach. In any case, conflict analysis should carefully analyse the BATNAs of the affected parties. On the other hand, where parties in a conflict have fought long and hard, but have reached a stalemate, the time may be ripe for a more collaborative approach that works to convince the parties that it is in their own self-interest to participate in a mediated discussion.

11.1. Conflict Assessment

A first step in conflict management is conflict assessment. An analysis of a particular conflict can provide insights into the nature, scope and stage of conflict and the approach(es) for its management. There are four main factors that need to be analysed in determining the scope, nature and stage of a conflict:

- *Characterization of conflict and stakeholders.* The type of conflict encountered, the number of stakeholders, and the relationships among them. The nature and origin of conflict, as well as the balance of power among the parties are analysed.
- *Stage in the project cycle.* Conflicts at the 'beginnings' stage are likely to be different than conflicts at the implementation stage. New stakeholders may arise as the project proceeds. This requires that the project be flexible and adaptive to changing circumstances.
- *Stage in the conflict process.* A determination of whether conflict is at a point at which interventions may be accepted.
- *Legal and institutional context.* The formal and informal institutions and the manner in which conflicts are resolved through these institutions and the formal legal doctrines may influence the appropriate approach.

Warner (2001) identifies five responses of people to conflict depending on the importance of achieving a goal or maintaining personal relationships:

- *Accommodation* – when one party wants to maintain personal relationships with the other party, he or she may choose to accommodate the other party's goal.
- *Withdrawal* – one party may opt to avoid confrontation or withdraw from the conflict because he or she is neither interested in maintaining a personal relationship nor concerned with achieving a goal. Withdrawal can often persuade reluctant and more powerful parties to negotiate towards consensus.
- *Force* – one party holds more power over another party and is not concerned

about damaging relationships and is keen on achieving the goal.

- *Compromise* – one party may have to give up something which results in a minimum 'win–lose' outcome.
- *Consensus* – involves avoiding tradeoffs and seeking a 'win–win' outcome.

11.2. Typology of Conflicts

Conflicts may arise due to different causes and at different levels which may suggest a certain approach to management (Boxes 11.2 and 11.3). Generally speaking, conflicts can be categorized into four groups based on the central critical situation or cause of the conflict:

- *Data and facts.* These types of conflicts can often be resolved by obtaining additional data, carrying out more studies, etc.
- *Needs and interests.* These conflicts may occur over sharing the benefits of projects, choices in the allocation of resources, or the financing of external costs. This type of conflict is the focus of most conflict management.
- *Values.* Conflicts over values, where values can be defined as deeply held beliefs, are usually not amenable to negotiation or other conflict management approaches. Here the solution may be to agree to disagree.
- *Relationships.* These are often caused by personality conflicts and may be resolved through mediation by a third party.

Box 11.2. Conflict in the Sokhulu Subsistence Mussel-harvesting Project, South Africa.

Prior to the project there was overt conflict between the conservation authority (NPB) and the community, sometimes manifested in violent interactions. This has largely ceased since the establishment of co-management and the subsistence zone. There is now, however, a measure of conflict between the joint co-management committee and the harvesters, centred on the limits imposed by the committee. The harvesters want to collect larger quantities per person. They also resent having to collect in relatively depleted areas when a 'healthy' stock exists in the adjacent control area. However, the Sokhulu members of the joint committee competently and confidently debate these management issues with the harvesters and stand their ground even in the face of angry harvesters. This conflict between community members of the committee and harvesters can be attributed largely to an increased understanding of sustainable management issues by Sokhulu members of the joint committee. Intermittent overt conflict exists between subsistence harvesters and recreational collectors. Although very few recreational collectors frequent the subsistence zone, some have been coming to the area for many years and resent being excluded. Confrontational individuals are, however, the exception and many recreational collectors react positively to the initiative when its aims are explained. Monitors are encouraged to talk to and inform recreational mussel collectors and to hand out a leaflet that explains the aims of the project.

Source: Harris *et al.* (2003).

Box 11.3. Conflict in the Soufriere Marine Management Area, St Lucia, Caribbean.

Conflict within the Soufriere Marine Management Area (SMMA) has taken place at the level of the Technical Advisory Committee (TAC) decision-making, not just among and between resource users. At various intervals, members of the TAC have felt that they have not had an equal say in decision-making and that certain parties have attempted to dominate the TAC. TAC members have, on occasion, also felt that the TAC was not always the place where decisions were made, for example, deals were negotiated outside the TAC. The recent declaration that the TAC is the forum for all SMMA decision-making may eliminate some of the conflicts in this area that have plagued the SMMA for years.

Source: Brown (1997).

Conflicts may be well-defined (sharp boundaries and constraints; clear solutions may exist) or ill-defined (unclear objectives and values; difficult to identify solutions). Relationships and the balance of power among the parties involved are important issues in all conflicts. Differing value systems may affect the relationship between the parties. Imbalances of power are not conducive to even-handed negotiation.

Fisheries and coastal management conflicts are usually multi-issue, multiparty conflicts, which adds to the complexity of dealing with them.

11.3. Approaches to Conflict Management

Conflict management is often used as the overarching term for both conflict prevention, or consensus-building approaches, and conflict resolution approaches. The latter approaches are often referred to as alternative dispute resolution (ADR). ADR refers to a variety of collaborative approaches, developed in North America, including conciliation, negotiation and mediation (Pendzich *et al.*, 1994; Moore, 1996). They differ in the extent to which the parties in conflict control the process and outcome. *Conciliation or arbitration* consists of an attempt by a neutral third party to communicate separately with disputing parties to reduce tensions and reach agreement on a process for addressing a dispute. The third party has legal authority to impose a solution. *Negotiation* is a voluntary process in which parties meet 'face-to-face', with or without the assistance of a facilitator, to reach a mutually acceptable resolution of the issues in a conflict. *Mediation* involves the assistance of a neutral third party, a mediator, who helps the parties in conflict jointly reach agreement in a negotiation process but has no power to direct the parties or enforce a solution in a dispute. As with negotiation, this is a voluntary process. Through ADR, multiparty 'win–win' options are sought by focusing on the problem (not the person) and by creating awareness of interdependence among stakeholders (Buckles and Rusnak, 1999).

There is some question whether or not ADR methods work in conflicts involving natural resources (Buckles and Rusnak, 1999). Techniques of ADR are dependent upon specific cultural, institutional and legal conditions, such as volunteerism, willingness to publicly acknowledge a conflict, and administrative and financial support for negotiated solutions, which may not be present in every context. Attitudes towards compromise, consensus or mediation vary. In some societies, openly discussing conflict may involve 'losing face'. ADR may be counterproductive if the process only manages to get certain groups together to mediate their differences when the causes of conflict and obstacles to resolution are beyond their control. There is also concern that there may develop a dependence on mediators to resolve conflict, to the neglect of building local capacity to do so. In addition, there is a need to acknowledge that people may use other mechanisms, such as peer pressure, ostracism, public humiliation and spirits, to resolve disputes. Chevalier and Buckles (1999) state that Western approaches to conflict management need to be balanced with the systematic study of local practices, insights and resources used to manage conflict.

Buckles and Rusnak (1999) state that *multistakeholder analysis* of problem areas and conflicts may also serve as an approach to conflict management that can address the complex interactions between stakeholders and natural resources at various levels. Multistakeholder analysis is a general analytical framework for examining the differences in interests and power relations among stakeholders, with a view to identifying who is affected by and who can influence current patterns of natural resource management and of *consensus building*. Various methods such as participatory rural appraisal, participatory research, stakeholder analysis, class and power analysis, and gender analysis can be used. Problem analysis from the points of view of all stakeholders can help to separate the multiple causes of conflict and bring a wealth of knowledge to bear on the identification and development of solutions. When stakeholders come to recognize for themselves the common interests and strategic differences that connect them to each other, new opportunities can emerge for turning conflict into collaboration. This approach is especially appropriate in early, strategic stages of the planning process, to develop directions or strategies that are supported by a large number of stakeholders.

11.4. Selecting an Approach

The approaches to conflict management range from multistakeholder analysis and consensus building (with the objective of fostering productive communication and collaboration prior to the outbreak of conflict by employing tools such as conflict anticipation and collaborative planning, together with the cultivation of alliances and mobilization of support) to managing conflict through negotiation, mediation and arbitration where the objective is to address conflict after it has erupted.

Conflict is a dynamic process that generally progresses from initiation to escalation, controlled maintenance, abatement and termination/resolution. There are generally four stages to every conflict, with appropriate approaches to management:

- Potential or dormant conflicts (consensus building/relationship building);
- Erupting conflict, with positions being developed (range of options, depending on the nature of conflict and relationship among parties);
- Evolving conflict, evolving towards a stalemate (mediation or arbitration) or evolving towards resolution/abatement (no assistance or facilitation);
- Resolved conflicts (depends on situation).

Choosing the correct approach through which to address a particular conflict is in itself a strategic choice (Box 11.4). Parties to a dispute must first decide whether to seek resolution to a conflict through a non-consensual process or through a more collaborative means. Once the decision has been made to use alternative conflict management processes, the parties must decide on which specific approach to employ. No single approach is effective in all cases. The circumstances of conflict and therefore the obstacles to agreement vary from one case to another. Disputes may involve many or few parties, the problem may be more or less urgent, emotional investment of the stakeholders may vary, the public interest may or may not be at stake, and the factors involved may be well understood or may be uncertain. Gaining expertise in conflict management includes learning about the specific advantages and disadvantages of the various approaches, and assessing which one is best in addressing a particular conflict situation.

Box 11.4. Tools and Techniques for Conflict Management.

- Several team-building tools are often used to build consensus and/or resolve conflicts.
- In some situations, Memorandum of Agreement (MOA) and organizational bylaws can be effective tools for preventing conflicts. By clearly stipulating the roles and responsibilities of group members, organizational bylaws concretely lay out members' expectations of one another. All partners can sign a MOA that details the roles, responsibilities and ownership of results.
- Laying down 'rules of the game' is also important in resolving conflicts. Rules should be few and simple. In one case, only two rules were put forward to resolve conflict: (i) an issue should always be discussed with a possible solution; and (ii) a possible solution can only be put forward if the person making the suggestion is committed to applying the solution.
- Games might also be an effective tool for developing the skills to work cooperatively. A game called 'Chasing the Dragon' can be used during team-building sessions. In this game, two groups form two lines and the person in the front is the head of the dragon while the ones in the back become the tail. The objective of the game is for the head of each group to chase the tail of the other group. The idea is for each group (or dragon) to protect its tail by following the movement of the head and moving in a concerted manner. After playing the game, the facilitator asks the groups about their feelings and learning in playing the game. Most likely, the participants' responses will revolve around the issues on the importance of working collectively. Games like this can then be used as starting points for discussing the relevant factors in collective work.

Source: Brzeski *et al.* (2001, pp. 119–120).

It is important to recognize that, although there are considerable differences between approaches that can be employed, there are also significant overlaps. Most approaches will involve some element of relationship building, procedural assistance, and possibly substantive assistance or advice as well. The use of conflict prevention, or consensus building approaches on one side, does not imply that there have not yet been conflicts between the parties. Similarly, the use of arbitrage, on the other side, does not imply that it will be more effective if the arbitrator manages to get the parties to cooperate as much as possible.

Borrini-Feyerabend (1997) states that to avoid focusing on particular stakeholders or positions (either of which can increase conflict and/or result in a deadlock), the best approach to adopt is what is sometimes termed 'interest-based' or 'principled' negotiation/mediation. This approach requires the parties to acknowledge that, to be sustainable, an agreement must meet as many of their mutual and complimentary interests as possible. The focus should be on mutual cooperation rather than unwilling compromise. This approach encompasses four general principles:

- Focus on underlying interests. When all interests (people's needs and concerns) are all satisfied it will be much more likely to result in a lasting and satisfactory resolution than one where the interests of only one side are addressed. Compromise may best serve everyone's interests.
- Address both the procedural and substantive dimensions of the conflict. Both the need to be included in decision-making and have opinions heard and to have interests addressed are met.
- Include all significantly affected stakeholders in arriving at a solution. Failure to include all stakeholders may lead to unsustainable solutions and new conflicts.
- Understand the power that various stakeholders have, and take that into account in the process. Each party's approach to the conflict will depend on their view of the power they have in relation to the other stakeholders.

11.5. A Process of Conflict Management

The Forestry Policy and Planning Division of FAO, in close collaboration with the Regional Community Forestry Training Center in Bangkok, Thailand, has developed a comprehensive training package on Community-based Forest Resource Conflict Management (2002). While focused on forestry, the process is also relevant to conflict management in fisheries and coastal resources. While there is a great deal of detail in the publication, only the map of the process will be presented here. The process involves nine steps and four key elements supporting the process.

1. *Entry point* – A process for managing a conflict may be initiated by any of the stakeholders, those who are directly involved (fishers) or those who are more distant (NGO or government). Stakeholders can act on the conflict at any stage (latent, emerging or manifest).

2. *Preliminary analysis of conflict* – The initiating stakeholders undertake this analysis of conflict to determine who needs to be involved, and the scale and boundaries of the conflict. Define the conflict issue, explore the causes, analyse stakeholder interests and needs, identify positions adopted, identify power they have, and identify incentives and disincentives to resolve conflict. This sets out an initial strategy for addressing the conflict that they can modify and develop with the input of other stakeholders.

3. *Broader engagement of stakeholders* – Attempts are made to engage the other stakeholders identified in the preliminary conflict analysis. Obtaining their interest and willingness to participate may require one or more actions, including shuttle mediation, raising public awareness about the conflict management effort, sharing with them the preliminary analysis of the conflict, etc. At this point, the capacity of stakeholders to engage in negotiation, if not already present, should be developed.

4. *Stakeholder analysis of conflict* – Individual stakeholder groups need to carry out their own analysis of the conflict. The level of detail of this analysis and timeframe will vary depending on the intensity, scale and stage of conflict. This analysis may identify other key stakeholders or stakeholder groups to involve and engaging in identifying key issues, etc.

5. *Assessment of conflict management options* – The analysis of conflict assists the stakeholders to assess, weigh and expand the various options available for managing or intensifying the conflict. Stakeholders evaluate then select what they believe is the best response and strategy for achieving their interests. Stakeholder actions could include: withdrawal, use of force, accommodating other groups' interests, compromise or collaboration. Doing nothing and taking a 'wait and see' approach may also be a chosen strategy. In deciding on the most suitable set of actions, stakeholders consider their options in terms of possible outcomes or impacts, the likely choices of other stakeholders, power imbalances, and differences in stakeholder capacity. Participatory action research methods can be used. Stakeholders also need to consider the context in which they will act – the appropriateness of using traditional approaches to managing conflict, available legal or administrative measures, or the desirability of mediation or facilitation for negotiations.

6. *Agreement on strategy for managing conflict* – All stakeholders enter into this process. They must agree on the guidelines for the process and what actions and capacities are required to support it. If they have not already done so, they will need to decide whether or not to use a third party, including the role and responsibility of the third party.

7. *Negotiation of agreements* – Stakeholders negotiate agreements based on the individual and shared needs and interests they have identified. They seek mutual gain agreements. Often agreements are made progressively and incrementally. As one agreement is implemented successfully it demonstrates commitment of the parties, this then provides greater trust upon which to build further agreements. With each agreement, stakeholders decide how they will implement and monitor it. In the process of negotiations and in deciding how they will implement and monitor the agreement, other stakeholders may be identified that need to be included in managing conflict. Similarly, stakeholders may discover new information needs and take action to obtain this information before they can make further agreements.

8. *Implementation of agreements* – The stakeholders implement and monitor agreements as they are made. Agreements are continually monitored, informing the various parties if they should continue to proceed or modify their strategy or decisions.

9. *Evaluation, learning and conflict anticipation* – Stakeholders evaluate the outcomes and impacts of conflict and the process of managing the conflict. This can occur at pre-determined points specifically set for evaluating conflict. This may also occur at points within the broader and overall co-management process. These evaluations aim to increase stakeholder learning and identify necessary changes to support improved collaboration in co-management. By evaluating outcomes, stakeholders can determine how to improve methods and systems for anticipating further conflicts. Setting achievable benchmarks or tasks by which to judge progress is an effective way to maintain the motivation of stakeholders. When agreements encounter difficulty or fail, stakeholders may need to revisit the agreements, obtain missing information, identify additional stakeholders, or identify other solutions in support of co-management. A conflict management process can therefore be an iterative or continuous cycle that adapts to new questions and changing needs.

Key elements supporting the process:

10. *Information needs and management* – The availability, management and acceptance of information are significant issues in managing conflict. Information plays a pivotal role in understanding conflict and the details of interests, clarifying shared goals and assessing the feasibility of solutions. The process outlines must allow time for stakeholders to check for and explicitly address information needs and issues. Stakeholders must agree to the information needs, the sources of valid information and how information will be exchanged and disseminated.

11. *Capacity building* – Building sustainable solutions for managing conflict requires evaluating not only the interests of stakeholders but also their capacity to participate effectively in the process. Developing capacity can vary in scale and focus from strengthening institutions and organizations to centring on the needs of specific individuals. Addressing conflict embraces a range of capacities – knowledge, skills, attitudes, organizational structures and logistical support. This is the same set of capacities needed for effective co-management.

12. *Consensus-based decision-making* – Consensus building aims to generate agreements and outcomes that are acceptable to all stakeholders with a minimum compromise. The intention is to find win–win solutions. In this process stakeholders are encouraged to identify and then meet underlying needs of all parties and be creative in the solutions they explore. Acknowledging the perceptions of others, ensuring good communications, building rapport and trust, and striving to continually widen options are key ingredients in this process.

13. *Keeping people informed* – Discussions between stakeholders are often carried out by representatives of key groups. An important part of the conflict management process is establishing reliable communication between stakeholder representatives and their constituency so that all who are involved and impacted by the conflict are informed and able to provide meaningful input.

Borrini-Feyerabend (1997) identifies steps in the process of negotiation:

1. Set a time and place to meet that is agreeable to all parties.
2. At the beginning of the negotiation, ask each party to explain their position clearly: what they want and why. They should not be interrupted except for points of clarification.
3. After all parties have stated their case, identify where there are areas of agreement.
4. Identify any additional information that any of the parties need in order for them to be able to understand the claims made by other parties. If necessary, stop the process until they can be provided with that information.
5. Identify areas of disagreement.
6. Agree on a common overall goal for the negotiation.
7. Help the parties to compile a list of possible options to meet this goal.
8. List criteria against which each option should be measured.
9. Evaluate each option against the criteria.
10. Facilitate an agreement on one or more options that maximize mutual satisfaction among the parties.
11. Decide on the processes, responsibilities and time frames for any actions required to implement the agreement.
12. Write up any decisions reached and get the parties to sign their agreement.

11.6. Conditions for Conflict Management

There are a number of conditions that can affect the success of conflict management. These include (Borrini-Feyerabend, 1997; Rijsberman, 1999; VSO, nd):

- The balance of power among stakeholders must not be uneven.
- The balance of power can be levelled through such methods as policy advocacy, community empowerment, networking to increase the power of marginalized groups, public education, building trust with outside groups.
- An independent mediator/facilitator can have a neutralizing effect on power imbalances by ensuring fair rules of discussion and negotiation.
- All parties must be willing to participate and settle their disagreements.
- All parties should be ready (skills, information) to negotiate.
- Alternative approaches should be available if the negotiation fails.
- Build personal relationships.
- Start with small issues that are easily settled.

- Establish process ground rules that are likely to create trust among the stakeholders.
- Each party should have some means of influencing the attitudes and/or behaviour of the other parties if they are to reach an agreement on issues over which they disagree.
- The parties should be dependent on each other to have their needs met or interests satisfied.
- The parties should be willing to compromise even though this may not be necessary.
- Representatives should have the authority to make decisions.
- The agreement should be feasible and the parties should be able to put it into action.
- The parties should feel some urgency to reach a decision.

Conflict management entails risks, such as (Rijsberman, 1999):

- It can be very demanding of time, money and other resources.
- Powerful stakeholders may capture the process and use it to coerce others to their position.
- Stakeholders may not have the skills to engage effectively in negotiation.
- Latent disputes may surface.
- It cannot resolve all issues. Conflict management can only ensure that alternative options are explored and that the parties in a conflict do not persist in holding differences that are the result of misunderstanding and that have no basis in reality.

12 Co-management Plan Implementation

IDRC, P. Bennett.

Once the plans and agreements have been approved and celebrated, the implementation should be started as soon as possible in order to capitalize on the good will and excitement generated by the negotiations (Borrini-Feyerabend *et al.*, 2000). Implementation comprises the activities by which the co-management plan is carried out. The implementation process will involve numerous decision-making points and a different process from the one used to create the plan and agreements (Table 12.1). All the activities in the co-management plan must be implemented correctly and in a timely manner if goal and objectives are to be achieved.

Table 12.1. Co-management plan implementation.

Stakeholder	Role
Fisher/fisher organization	• Participate in implementation • Comply with rules and regulations • Participate in monitoring • Attend education and training activities • Provide input into development and livelihood activities • Provide information and feedback on programme
Other stakeholders	• Participate in implementation • Comply with rules and regulations • Provide information and feedback on programme
Government	• Support implementation of plan • Institutionalize programme support • Establish co-management implementing structure • Draft and endorse ordinances • Assist in revenue collection and support
External agent/CO	• Build capacity of co-management organization to obtain funding • Provide technical assistance and training • Ensure participation of organizations and other stakeholders in programme • Strengthen community organizations • Develop and train leaders • Assist in monitoring • Assist in community development and livelihood activities

> Coastal management programs are not self-executing. Implementation involves not just the daily planning, regulatory enforcement decisions and activities undertaken by government officials, but the decisions and activities of NGOs, resource user groups, and community residents. For effectiveness and legitimacy of the coastal management program, the government and non-government stakeholders must perform their assigned roles in carrying out the plan. Having a plan is not enough; follow through and commitment to fully implement the plan are also required.
>
> (DENR *et al.*, 2001b, p. 55)

It is common to feel overwhelmed by implementation. There is so much to do and so much to coordinate. Implementation will require trusting the plan and trusting the partners and staff. No plan is perfect. There will be successes and failures. This is why continual monitoring and learning-by-doing has been emphasized. There may be failures early on as everyone learns to work together and do their job, by learning from it and moving forward.

While implementation is based on the plan and agreements, the quality and effectiveness of implementation are shaped by a number of factors (National Roundtable on the Environment and the Economy, 1998; DENR *et al.*, 2001b):

• A collaborative decision-making process;
• Legal authority to manage;

- Adequate and dedicated resources (personnel, funding, equipment) for management;
- Information and data to support monitoring and learning-by-doing;
- Key political support;
- Staff skills and commitment;
- Coordination arrangements with government, external agents, resource user groups and community members;
- Community support through participatory processes.

Please note that no one component of implementation exists in isolation, but they are linked and complementary to each other. For example, solutions to fisheries management problems often lie outside the fishery, thus requiring household livelihood and community development issues to be addressed.

12.1. Setting to Work

The workplan should specify an individual or group who will be responsible for each activity and for reporting on the progress being made to the stakeholders.

Especially in the early period of implementation, there will emerge various interpretations of the agreements and rules. For more formal agreements, the contracts and law will provide some basic reference. For less formal agreements, conflict management processes will need to be followed. An important principle to be applied is that of accountability. *Accountability* is the clear and transparent assumption of responsibilities, the capacity and willingness to respond about one's own actions (or inactions), and the acceptance of relevant consequences (Borrini-Feyerabend *et al.*, 2000). It is important that the co-management programme remains flexible and creative to changing human and natural needs and conditions.

While the plans and agreements are being implemented, the fishers using the resource tend to develop a greater sense of the legitimacy of their role and responsibility. This may encourage them to refine the management measures and rules and to apply more efficient and complex solutions. The scope and scale covered by the plan and agreements (new community organizations, other fishing gear types, new boundaries) may also increase in size. The co-management organization will need to be innovative in adapting to these changes. This is facilitated by some experimentation, flexible management plans and budgets, and by having sensible people in charge (Borrini-Feyerabend *et al.*, 2000).

12.2. Management Measures

While it is beyond the scope of this handbook to fully discuss all the fisheries resource management measures available, a short discussion will be presented as found in Berkes *et al.* (2001, Chapter 6).

12.2.1. Traditional measures

Documented cases have accumulated, especially since the 1980s, on long-standing community-based management systems. It is becoming clear that these time-tested systems were often based on sound ecological knowledge and understanding, particularly in the Asia-Pacific region, which is rich in traditional knowledge and management systems. Many of these systems have been documented in detail, especially those in Japan and parts of Oceania (Melanesia, Micronesia and Polynesia) (Ruddle and Akimichi, 1984; Freeman *et al.*, 1991).

The most widespread single marine conservation measure employed in Oceania was a combination of reef and lagoon tenure and taboos. The basic idea behind reef and lagoon tenure is self-interest and sustainability. The right to harvest the resources of a particular area was controlled by a social group, such as a family or clan (or a chief acting on behalf of the group), who thus regulated the exploitation of their own marine resources. A wide range of traditional regulations and restrictions applied to resource use. Some of these rules could be attributed to religious beliefs (Johannes, 1978) and some to power relationships and regional differences in systems of political authority (Chapman, 1987). But by and large, reef and lagoon tenure rules served both conflict resolution and conservation, directly or indirectly, and operated as institutions for the management of common property resources.

12.2.2. Conventional and new fisheries management

Much has been written on tools or control measures for fishery management. Some were developed out of common knowledge and long-term observation in traditional management systems (Table 12.2). These and others developed in the last 100 years as part of conventional fisheries management, such as

Table 12.2. Frequency of occurrence of traditional fishing regulations in 32 societies from around the world. (Source: Wilson *et al.*, 1994.)

Type of regulation	Frequency
Areas (community controlled)	30
Limited access	16
Technology	12
Seasonal limits	10
Protect breeding stock	9
Protect young	8
Conservation ethic (of individuals)	6
Size limits	4
Overcrowding	3
Quotas (on catch)	1
Other	8

those described by Beddington and Rettig (1983) and Hilborn and Walters (1992), comprise the present tool box for fisheries managers (Table 12.3).

The main difference between the old and new approaches is the way of reaching decisions on when and how to apply the tools.

12.2.3. Rights-based fisheries

Economists have pointed out that governments often use the wrong approach to deal with problems that arise from a 'market failure'. Command-and-control

Table 12.3. An overview of the main methods used for control of fisheries.

Method	Aim/effect	Comments
Licensing, limited entry (ID)	Licensing is the only way to directly limit the number of participants in the fishery.	Licences can be used as a means for recovering some revenue from the fisheries. Licensing alone is seldom enough to control the amount of fishing effort.
Effort limits (ID)	These direct limits to the number of units of effort; for example, hours fished, traps pulled or trawl sets.	Limiting effort in this way is more direct, but fishers usually find ways of getting around effort limits by increasing aspects of effort that are not limited, i.e. larger traps or larger boats.
Closed season (II), closed area (II)	These aim to protect a specific part of the stock known to occur in or at a particular place or time; usually spawning or young fish. May also be used to control total effort by eliminating fishing from a particular area of the stock or period of the year.	When used as a means of controlling total effort, fishing usually increases in the open area and at the open time of the year. Thus, reduction of effort is not directly proportional to the closed season or area. Closed seasons are easier to monitor than closed areas, unless the latter are very large.
Gear restrictions (II)	These usually aim to control the size or species of fish caught; for example, by regulating the mesh size used in nets or traps.	Although the relationship between gear and size of fish caught is imprecise, gear restrictions can be monitored by inspection ashore.
Catch quotas, total allowable catch (TAC) (OD)	Quotas directly limit the amount of fish taken from the stock to that corresponding to the target reference point. TAC is the simplest form of catch quota.	Catch quotas vary with the abundance of the resource and must thus be re-estimated at regular intervals. This requires substantial amounts of detailed data. Regulation by catch quotas also requires that fish landings be monitored on a real-time basis so that the fishery can be closed by the catch when the quota has been taken. A single TAC often results in a race for the quota and, consequently, overcapitalization.

Continued

Table 12.3. *Continued.*

Method	Aim/effect	Comments
Industry quotas (ID)	The TAC is divided up among participants in the fishery.	Individual fishing companies can manage the way in which they take their share in order to optimize their economic return. The equitable distribution of quotas among participants is usually difficult and contentious.
Individual transferable quotas (ITQs) (OD)	This is a form of industry quota in which the quotas may be transferred, sold or traded.	ITQs facilitate the operation of normal market effects in the fishing industry. More efficient companies can buy quotas and so increase their share of the resource. A basic proportion of ITQs are given out on a long-term basis so that companies may plan their operations. Remaining quotas are distributed or sold each year, with the amount becoming available being dependent on the abundance of the resource. May lead to monopolies.
Size limits (OD)	This directly limits the size of fish landed in order to reduce growth of overfishing and to ensure that immature individuals are not caught.	Shore-based monitoring of size limits will often lead to discarding of smaller sizes at sea. Because discarded individuals usually die, this defeats the purpose of the regulation.
Taxes or tariffs (OI)	Taxation on the fish landed is one means of reducing the amount of fish caught.	This increases the cost of fishing, thus shifting the cost and revenue equilibrium toward lower effort.

I/O, input/output control; D/I, direct/indirect control.

regulations used in an attempt to correct the problem through legal and administrative means do not dissolve the market failure. This has led to a market-based perspective that suggests that governments should change the incentive structure in order to bring private and public interests closer in line, restoring the workings of a 'perfect' market. Advocates of a market-based perspective see the lack of individual property rights as a market imperfection in the fisheries.

This perspective led to the development of property rights regimes to regulate access and effort. In temperate developed countries, biological fisheries management used biological knowledge of fish stocks to set Total Allowable Catches (TAC) designed to restrict exploitation of the stock at or below the TAC (Beddington and Rettig, 1983). Economic fisheries management included economic instruments to regulate exploitation at the TAC, thereby extracting maximum economic rent. The introduction of Individual Quotas (IQs) and Individual Transferable Quotas (ITQs) were steps towards individual property rights that, ideally, help to restore the workings of the market mechanism, address problems of governance, and add objectivity in adjusting fishing effort.

These management models, however, seem better suited to temperate regions with discrete single-species fisheries, and therefore calculable TAC, than to the multi-species, multi-gear fisheries of many tropical countries. These models have limited applicability to tropical fisheries because of the large amount of information that managers need to implement them, the wide variety of fishing gears used in the tropics, and managers' limited ability to control access of fishers, both full- and part-time, to the tropical fishery.

Quotas, particularly individual transferable ones, have been promoted as appropriate rights-based management tools for several fisheries in developed countries. They are also being introduced to some developing countries, but there are several reasons why quotas are problematic, especially for small-scale fisheries (Table 12.4). However, other tools based on fishing rights have been used for centuries in the traditional fisheries of many communities, particularly in the small islands of the Pacific.

Several of the constraints and challenges facing small-scale fisheries managers have been identified before, and Table 12.4 relates these to quota systems. Several of them are also problems for large-scale and developed-country fisheries, but in these cases the nature of either the fisheries resources or management capacity makes them less critical.

Other rights-based management tools have focused on quotas for groups

Table 12.4. Reasons why quotas are problematic for small-scale fisheries. (Source: Copes 1986.)

Quota and fishery features	Issues that may confront many small-scale fisheries
Quota busting	Poor enforcement resulting in quotas often being exceeded
Data fouling	Inaccurate catch reporting due to cheating or complexity
High variation stocks	Widely variable year classes, abundance, availability, etc.
Short-lived species	No clear relation between stock and next year's recruitment
Flash fisheries	Season too short to be monitored for management
Real-time management	Precise control of effort difficult with dispersed fisheries
High-grading	Market strategy of discarding low-value fish encouraged
Multispecies fisheries	Not possible to set optimal catch or effort for a complex of species
In-season variation	Declining abundance in-season resulting in a race at the start
Information for TAC setting	Information base inadequate for setting the TAC with precision
Transitional gains trap	Unpopularity of taxing the gains of initial beneficiaries
Industry acceptance	Low acceptance if initial allocations are seen as inequitable
Spatial distribution of effort	Overexploitation of high-yielding grounds due to patchiness
Quota concentration	A few companies or rich people buying out many small fishers
Social and economic change	Affecting society more than many other management tools

such as communities or fishing-industry organizations rather than individuals or companies. The most durable systems have concerned access to the fishing area or gear rather than to the amount of catch. The common feature is the decentralization of control over exploitation, usually by devolving power that was centred upon the state. In the past, this control was often community-centred and at the origin of the rights-based system integrated with the social and cultural practices of the resource users. Now, formal recognition by the state is also necessary for full legitimization and acceptance by the wider society and by outsiders who may seek to impose different values.

12.2.4. Ecosystem-based measures

The impact of environmental degradation from both fishery and non-fishery activities on the ecosystems that support fisheries, particularly inland, coastal, and inshore fisheries, is increasingly recognized as the major fisheries management problem (Dayton *et al.*, 1995). Separating these impacts on exploited resources from the direct effects of fishing mortality may be one of the major challenges of fisheries management planning. Since most small-scale fisheries are near shore, non-fishery human impact is usually a more important issue in their management than in large-scale fisheries. Consequently, different types of management measures are likely to be useful, depending on distance from shore (Caddy, 1999). For inshore and inland fisheries, habitat conservation, rehabilitation and enhancement are commonly used management measures.

Although this is an emerging field, it appears that ecosystem-based measures will be variations of standard measures based on ecosystem criteria. For example, areas may be closed to protect habitats; quotas of prey species may be set to ensure adequate forage for predators; and predator quotas may be set to ensure that predator depletion does not lead to explosions of prey populations that are released from predation pressure.

Marine protected areas

Marine protected areas (MPAs), such as reserves, sanctuaries and parks, can achieve protection of well-defined areas and critical habitats, such as coral reefs, mangrove, seagrass and wetlands (Box 12.1). When properly sited and designed, an MPA can play a significant role in aquatic resource conservation. MPAs help to sustain and increase biotic and genetic diversity by protecting rare, threatened and endangered species, subpopulations and their habitats. By restricting fish harvest, they give different species the chance to freely reproduce. As fish inside the MPA grow larger and multiply more easily, this leads to a faster turnover of fish from the MPA to the non-MPA area, which increases yields for fishers. An MPA can also reduce conflicts between fishers and other users by providing areas where non-fishery users can pursue non-consumptive uses of the resources.

One of the main concerns about relying too much on MPAs is that they simply displace fishing into adjacent areas, leading to extra depletion there

---

Box 12.1. MPA Terminology.

The following terms can be applied to MPAs:

- *Marine protected area* – Any specific marine area which has been reserved by law or other effective means and is governed by specific rules or guidelines to manage activities and protect part or the entire enclosed coastal and marine environment.
- *Sanctuary* – An MPA where all extractive practices, such as fishing, shell collection, seaweed gleaning, and collecting of anything else is prohibited. It also allows for control of other human activities, including access, in order to protect the ecosystem within the specific site.
- *Reserve* – An MPA where strict sanctuary conditions are not mandated for the entire area yet there is still a desire to control access and activities, such as boating, mooring and various fishing techniques. It allows for zones that include a sanctuary area.
- *Marine park* – An MPA where multiple uses are encouraged that emphasize education, recreation and preservation; usually implemented by zonation schemes that can include a sanctuary area.

Source: White (1988).

(Fogarty, 1999). This, in turn, increases the difference in abundance between protected and exploited areas, which may increase emigration of fish from the former to the latter, thus reducing abundance in the protected areas. Another concern is that strict protection through MPAs can create conflicts among different interests, user groups, levels of government and national government agencies.

One of the reasons that parks frequently do not perform as intended is that many are set up with the hope that they will serve several purposes: tourism, biodiversity conservation and fisheries enhancement; purposes that are not always compatible. Consequently, one must pay careful attention to site selection and design criteria when establishing MPAs.

Success of MPAs will depend on managers being in tune with fishers and committed to working with fishing communities. MPAs that are to protect the viability of major fish populations and critical habitats need to be designed with those objectives in mind, not to carry out the requirements for arbitrarily designated protected areas. They will need the local fishing communities' support and help in siting, design, management, monitoring and enforcement.

MPAs can be one important management strategy within a larger area-wide coastal management framework whose broader goals may include maintaining essential ecological processes and life support systems, maintaining genetic diversity and ensuring sustainable utilization of species and ecosystems. MPAs are closely linked to issues of ownership and control over specific pieces of coastal marine space. Their management outcomes greatly affect activities that degrade local coastal conditions.

12.2.5. Habitat restoration, creation and enhancement

Small-scale fisheries, usually located in inland water bodies and coastal areas, are highly dependent on habitats (coral reefs, mangroves, seagrass, wetlands) that are susceptible to human-caused pollution and physical destruction. The restoration of these habitats, particularly those that limit the abundance of a resource at some life-history stage, may be the most important step to increasing stock productivity. The restoration of coastal habitats that have been destroyed by development is increasingly taking place in many developed countries. In some cases, lands that have been filled and reclaimed for agriculture and development have been purchased, at very high cost, and returned, insofar as is possible, to their original condition. This trend is based on the realization that many of these habitats are important nursery or spawning areas for fishery resource species. The role of coastal wetlands in maintaining the quality of fresh water that is discharged into nearshore habitats, and thus the integrity of these ecosystems, has been another driving force in coastal wetland rehabilitation.

A number of methods exist to create and enhance aquatic systems. Examples of these are artificial reefs, fish aggregating devices for pelagics, and casitas for lobsters. Many of these have traditionally been used by small-scale fishers around the world (Kapetsky, 1985). (A review of these approaches, which are adequately covered in other publications, is beyond the scope of this book.) However, before implementing them, the small-scale fishery manager must consider whether they are likely to contribute to increased production by the resource or simply increase the availability of the resource to exploitation by aggregating individuals. If the latter, their use must be accompanied by the capability to control exploitation.

Artificial reefs (ARs) are structures that serve as shelter and habitat, source of food, breeding area, resource management tool and shoreline protection. The AR may act as an aggregating device to existing dispersed organisms in the area and/or allow secondary biomass production through increased survival and growth of new individuals by providing new or additional habitat space. In addition, ARs have been considered as a barrier to limit trawling in coastal areas where they may be in conflict with small-scale fishers.

Fish aggregating devices (FADs), items placed in the water to attract fish to aggregate (gather near to them), have been used in Southeast Asia for much of the 20th century, if not longer. FADs are deployed in a variety of environments, from calm waters to rough, high-energy environments (Pollnac and Poggie, 1997). They can be constructed from a wide range of materials, from simple line and palm fronds to sophisticated devices with radio beacons. For example, bamboo rafts are traditionally used in Japan. The benefits of using FADs include: (i) increased catch, (ii) lowered fuel consumption, (iii) accessibility to small-scale fishers, (iv) shifted effort from overfished areas, (v) improved fishing vessel safety, and (vi) definition of territory and/or inhibition of certain types of fishing. Potential problems include: (i) increased probability of stock depletion, (ii) changes in eating habits of attracted fish, (iii) lack of monitoring and evaluation, (iv) restricted access to the resource, (v) increased conflicts, (vi) periodic maintenance and replacement required, and (vii) cost of long-lived, high-technology devices (if used).

12.2.6. Restocking and introductions

Enhancing fish populations by restocking with young individuals has been most successful in small, enclosed water bodies such as ponds and lagoons (Welcomme, 1998). The generally high cost of producing the young for stocking means that this approach has been most cost-effective for recreational fisheries that provide economic returns beyond the landed value of the fish. The few instances of successful stocking programmes in the marine environment are in very localized inshore habitats (Blaxter, 2000). The costs and benefits of any stocking programme should be evaluated. As well, the variety of risks, such as of genetic dilution of the wild stocks and introduction of disease, should be considered.

When considering introducing new species, the manager should fully explore the extensive literature that describes the many pitfalls and case histories of unexpected consequences. The Code of Conduct for Responsible Fishing provides guidelines for introductions (FAO, 1996).

12.2.7. People-focused measures

Successful implementation of fisheries management is now seen to include a variety of measures that engage and inform stakeholders (including the public). Addressing the undesirable social and economic implications of attempts to reduce fishing effort is also an emerging direction.

In several parts of this handbook, the need to inform and build the capacity of fishing-industry stakeholders in order to empower them to participate in fishery development and management has been emphasized. This should be borne in mind as a crucial new direction for management.

12.2.8. Public education

Education aimed at the non-fishery public can increase their awareness that they are stakeholders with a right to expect that fisheries will be well managed on their behalf, and that industry stakeholders will observe the agreed-upon measures in return for the right, or privilege, to participate in the fishery. Messages absorbed by the general public, such as those about conservation, can be important for structuring social sanctions and attitudes. A knowledgeable public can also play a role in enforcement, either indirectly through the political directorate or directly by exercising its consumer right not to purchase illegally or inappropriately harvested products.

The role of the public as stakeholder should not be underestimated, particularly when household consumers are the primary purchasers of fish. If the public is aware of the issues, regulations and the long-term effects that breaking the regulations may have on the availability of the product, there is reason to believe that many individuals will choose not to purchase illegally caught fish. If properly informed and supported by the authorities, the public can also play a role in reporting violations. These roles may be strengthened through public education and market-oriented initiatives such as the eco-labelling mentioned earlier. Cases involving sea turtles and marine mammals are well known. However, public perception of the fairness of management also takes into account the opportunities (or lack thereof) for involved fishers to pursue alternative livelihoods.

12.2.9. Managing excess fishing capacity

It is now almost universally accepted that many coastal fisheries are overfished. Many small-scale fisheries are home to an excessive level of factor inputs (capital and labour) relative to that needed to catch available fish. Thus, most fisheries can be characterized as having the problem of 'excess capacity', 'overcapitalization' or simply 'too many fishers chasing too few fish'. The result is lower productivity of small-scale fisheries, increasing impoverishment of small-scale fishers, and erosion of food security in coastal communities that depend on fish supplies for protein and income.

Because the capital and labour employed in small-scale fisheries are generally use-specific, their exit is often difficult and painfully slow. As long as small-scale fishers can obtain a positive return, they will continue fishing, trying to circumvent any command-and-control regulatory measures such as gear limitations and closure of fishing areas. These measures appear to focus on the resource rather than on the people: the fishers, other resource stakeholders, and the community. Resource managers' action to deal with excess capacity as a major cause of resource overexploitation and environmental degradation reflects a one-sided policy response to the problem. Unless we address the core issue of excessive capacity, that is, by facilitating the exit of labour and capital from the fishery without unacceptably severe social and economic disruption, any regulatory measure or other management strategy will simply be a stopgap measure. People will continue to enter the fishery unless viable alternatives are presented.

As traditional institutions and methods of controlling overexploitation of fisheries fail under the pressures of modernization and market economies, fisheries managers are increasingly aware of the need to develop appropriate policies to facilitate the exit of capital and labour from overexploited fisheries. This growing consciousness of the importance of reducing fishing overcapacity culminated in the FAO Committee on Fisheries' adoption in February 1999 of the International Plan of Action for the Management of Fishing Capacity. This instrument calls for states to prepare and implement national plans to effectively manage fishing capacity, with priority to be given to managing capacity in fisheries where overfishing is known to exist. International policy discussions of the fishing fleet overcapacity problem have focused overwhelmingly on industrial fishing fleets, largely ignoring the problems of small-scale fisheries. Developing countries with small-scale fisheries with severe overcapacity are unlikely to prepare effective plans to address that aspect of fishing overcapacity without initiatives to help them analyse the problem and generate new policy options.

The problem of reducing excess capacity in small-scale fisheries in developing countries is much more complex than that of reducing overcapacity in industrial fleets. The complexity in small-scale fisheries is compounded by: growing populations, sluggish economies, fishers' high dependence on the resource for food and livelihood, a paucity of non-fishery employment, increasing numbers of part-time and seasonal fishers, limited transferability of and rigidities in the movement of use-specific capital and labour, and the lack of a coordinated and integrated approach to horizontal economic and community development that blends fishery and non-fishery sectors. Migration to the coast and the subsequent shift to fishing also occur because people in the countryside have lost their farms due to land conversion, or have avoided armed conflict situations in the upland areas.

Thus, a reduction of excess capacity implies an increased focus on people-related solutions and on communities. This should involve a broad programme of resource management and economic and community development that emphasizes access control and property rights, rural development, and linking of coastal communities to regional and national economic development. This new management direction needs to address coastal communities' challenges, including employment and income, food security, better quality of living, and delivery of community services. We must go beyond the 'common' solution, which is to give fishers 'pigs and chickens' as a supplemental livelihood, towards more innovative approaches involving development of skills and microenterprises and the use of information technology. Co-management and community-based natural resource management (CBNRM) strategies can provide a framework for such linked development and management initiatives. Community-centred co-management can serve as a mechanism not only for resource management but also for social, community and economic development by promoting participation and empowerment of people to solve problems and address community needs.

12.3. Community and Economic Development and Livelihoods

Occupational multiplicity, a prominent feature of small-scale fisheries, has the consequence that fishery management extends into the domain of integrated community and economic development, whether urban or rural. Many fisheries management measures alter patterns of employment in fisheries and supporting occupations. Community and economic development programmes that address alternative (replaces fishing as a source of income) and supplemental (adds to but does not replace fishing as a source of income) livelihood for fishers, and livelihood planning for fisher households and part-time fishers, can contribute to the management of fishing effort. Livelihood planning is especially important where households depend on fishery-related income.

A livelihood comprises the capabilities, assets (including material and social resources) and activities required for a means of living. Livelihood development focuses on increasing the capital to effectively set up and sustain viable and sustainable livelihoods. These include human capital (skills, knowledge), social capital (social resources such as networks and relationships), natural capital (natural resources), physical capital (basic infrastructure and producer goods) and financial capital (financial resources). Livelihood strategies in fisheries should focus on the whole household and all its members including husband, wife and children. The outputs of livelihood strategies are more income, increased well-being, reduced vulnerability, improved food security and a more sustainable natural resource base.

Livelihood projects can be classified as those that are resource-based (utilizing aquatic resources) and non-resource based (developed independent of aquatic resources). Projects may be income-generating projects that generate additional income for the household (such as pigs or mollusc aquaculture) or enterprises that are formed as businesses to generate more long-term jobs and benefits (such as handicraft enterprise or fish marketing). Entrepreneurship often requires the development of a broader skills and capability base for those involved. Any livelihood project, whether income generating or enterprise, should recognize cultural diversity, provide equal opportunity for both men and women, be economically viable with proper management, and environmentally friendly. A project feasibility study should be conducted before any livelihood is proposed. A project feasibility study is a thorough and systematic analysis of all factors (market, technical, financial, socio-economic and management) that affect the possibility of success of the proposed activity (TDC and CERD, 1998).

Extensive excursions into these areas are beyond the scope of this handbook, but they must be part of a new-style fisheries management. The capacity to contribute to such initiatives may require close links with agencies specifically responsible for community and economic development. Such development can also increase the quality of life in rural coastal communities through delivery of basic services (for example, health and education) and infrastructure development (for example, roads and communication).

Resource management policies must shift from a resource exploitation orientation to one of more holistic conservation and human resource management. In order for socio-economic development to be sustainable, attention must be given to policies that address issues of food security and people's well-being and livelihood, not just regulatory fisheries management.

Mixed with policies concerned with resource management and conservation is the need to address problems of poverty, unemployment and decreasing quality of life in fishing communities. The main brunt of such economic and social distress is borne by women, children and unskilled fishers, as well as by those unskilled people who depend, directly and indirectly, on the fishing industry. Elements of this prevailing scenario are: high levels of unemployment or underemployment, unavailable alternative or other supplemental employment and livelihood opportunities in the community, a growing population and pressure to find additional fisheries resources, lack of credit and markets, and the paucity of institutional mechanisms to undertake system-wide development.

12.4. Enforcement and Compliance

(The reference for this section is: Department of Environment and Natural Resources, Bureau of Fisheries and Aquatic Resources of the Department of Agriculture, Department of the Interior and Local Government, and Coastal Resource Management Project, 2001d; www.oneocean.org)

Regulations and ordinances must be enforced to protect fisheries resources. Compliance with the plan, agreements and rules is essential to the effectiveness of the whole co-management programme. There must be an enforcement mechanism that specifies who is responsible, the means of enforcement, and the penalties for non-compliance. While the national and local governments have responsibility for coastal law enforcement, enforcement of regulations by fishers is increasingly being considered by governments short of enforcement resources.

Law enforcement is more than the presence of armed police arresting people, it involves the application of a broad range of approaches by different institutions and stakeholders to change or modify behaviour. Oposa (1996) articulated four principles of effective environmental law implementation that can be applied to coastal law enforcement (DENR *et al.*, 2001d). These principles include:

1. Law is an agreement of minds or a 'social product' that must be deemed desirable and supported by a mental and emotional agreement by individuals and society at large.
2. Legal marketing or selling the law is necessary to promote voluntary compliance.
3. Sociocultural sensitivities and pressure points must be considered in the manner used for implementing the law.
4. Swift, painful and public punishment must be carried out in order to modify behaviour and serve as a deterrent.

Law enforcement can be viewed as a variety of interventions that government and the community may rely upon to achieve compliance with requirements of the law. These interventions can be 'soft' preventive measures such as public education or 'hard' sanctions imposed by apprehension, prosecution and conviction. Soft enforcement approaches promote voluntary compliance with the requirements of the law without going to the courts. Soft enforcement focuses on the social and cultural dynamics of compliance that can be used to: (i) sustain widespread compliance, (ii) encourage voluntary compliance, and (iii) achieve general deterrence. When widespread compliance is achieved, target sectors have an adequate level of knowledge and attitude on the issues and behave within the bounds of socially accepted practices and legal requirements. Soft or positive approaches include:

- Social marketing;
- Social mobilization;
- Coastal resource management best practices;
- Legislation and regulation;
- Information management and dissemination;
- Education and outreach;
- Monitoring and evaluation.

Negative or 'hard' enforcement uses legal sanctions imposed by a court or regulatory authority for deterrence. Hard enforcement approaches have one objective, that is to identify, locate and suppress the violator using all possible instruments of law. It involves the process of developing sophisticated strategies to apprehend repeat violators and negate all economic profits and benefits from illegal activities. Law enforcement activities are directed to a specific violator or violation. In these cases, the law enforcement approach to persistent violators must be swift, painful and public. Swift means it should be directed to a specific and pre-identified target. It should be painful financially and psychologically and it should be public so that other people will be aware of the consequences of the offences. Negative or hard approaches include:

- Continuous presence of law enforcers;
- Consistent activities to detect, apprehend and prosecute violators and impose appropriate sanctions;
- Sophisticated strategies developed to apprehend repeat violators;
- Negate all economic benefits from illegal activities.

Enforcement requires consultation and coordination among the various agencies and organizations with responsibility for enforcement about regulations, monitoring, surveillance, apprehension and sanctions. This includes the police, Navy, Coast Guard and any community-based enforcement units.

The inability to enforce, in the field, regulations that make perfect sense in the meeting room has been the downfall of many fisheries. Small-scale fisheries with large numbers of fishers widely dispersed in inaccessible places are particularly resistant to top-down enforcement. A host of factors come into play to make this type of enforcement ineffective. Small-scale fishers are often among the poorest people in society. Therefore, the political and judicial will to enforce regulations on them is often absent, especially when the action is seen as taking food from their family. In addition, the judicial systems are often bogged down with cases that the court inevitably perceive as more important than the enforcement of fishery regulations.

A community-based approach to enforcement may be warranted that involves the fishers in the regulatory and enforcement process. When the fishers understand the problems and benefits of taking action, and agree upon the actions to be taken, they will take part in the enforcement – at least to the extent of encouraging compliance. In a co-managed fishery, there is a greater moral obligation on individuals to comply with rules and regulations, since the fishers themselves are involved in formulating, rationalizing and imposing the rules and regulations for their overall well-being. The government will need to ensure that community-based enforcement units are trained and operational, with adequate equipment.

Fisher enforcement can take two forms that are not mutually exclusive. In the first form, fishers perform a mainly monitoring function, reporting violations to the authorities and exerting peer pressure. In the second form, fishers or other community members are legally designated as enforcement officers.

12.5. Monitoring

As previously discussed (see Chapter 10, Section 10.10), monitoring allows for an impact assessment of the co-management plan's activities in order to determine whether or not they are being achieved and what needs to be done to make improvements. A monitoring strategy is developed which includes indicators and the tasks needed to collect the data. The baseline data allow for a comparison of the results of the activity to be made against some benchmark. The comparisons that can be made in monitoring include:

* Comparing a group affected by the activity to itself over time by measuring how a given indicator changes as a result of the activity.
* Comparing a group affected by the activity to a group not affected by the activity over time by measuring how a given factor changes in a group affected by the activity relative to a similar group that is not influenced by the activity.

Both quantitative and qualitative data can be collected using a variety of methods. Participatory methods employ trained community members to collect and analyse data. Scientific surveys employ researchers in specialized fields of biology, social science and economics to collect and analyse data on biophysical and socio-economic indicators. Surveys and focus group discussions may be employed to assess attitudes and perceptions for behavioural indicators. The data collected should be systematically recorded and stored for easy access and use.

Analysis is a continual process of reviewing data as they are collected, classifying them, formulating additional questions, verifying data and drawing conclusions. Analysis is the process of making sense of the collected data. It should not be left until all the data have been collected (Margoluis and Salafsky, 1998). The results of the analysis are presented to both internal and external audiences.

The co-management plan is adapted based on the monitoring results. Iteration involves using the results of the monitoring to improve the plan. It may be found that activities are going as planned and little change is needed. However, it may also be found that things are not going as expected and big changes are needed to be made. This will require going back over the plan and its components to make modifications and move forward (Margoluis and Salafsky, 1998).

To facilitate learning-by-doing, a constructive attitude to both success and failure is required. If failures are regarded as an opportunity for learning, and if people are rewarded for identifying problems and promoting innovative solutions, learning-by-doing will be strongly encouraged. The challenge can be to recognize that plan adaptation and refinement is a normal activity that occurs because of experience and new information.

It should be noted that in carrying out the monitoring, it is necessary to document all the processes involved in order to develop institutional learning to help to avoid mistakes in the future.

12.6. Annual Evaluation, Workplan and Budgeting

On an annual basis, the co-management plan should be evaluated and a new workplan and budget developed. The evaluation should determine whether the context has changed, whether lessons have been learned from experience, and whether the co-management programme is on the right track. Evaluations are conducted to assess, among other purposes, the effectiveness of new plan strategies, diagnose implementation problems, make adjustments in strategies, and make decisions about plan management (see Chapter 6, Section 6.5.9.1). Monitoring serves to provide information for the annual evaluation. The results of the monitoring are presented to the co-management organization. The assumptions on which the activities are based are tested to determine which activities worked and which did not and why. Decisions about necessary modifications to the plan and activities are made and, if so, what changes are needed and who should carry them out. The process may involve negotiation and mediation, although generally at a faster pace than earlier. This information is provided as input into the annual workplan.

During annual workplan and budget preparation, the co-management organization, with other selected stakeholders, should hold a strategic planning workshop to identify activities, resource needs and funding requirements. All forms of revenue generation and financing of the plan should be reviewed and evaluated. Additional funding requirements should be identified.

Participatory monitoring and evaluation is a potent source of information for feedbacks, which in turn, are sources of energy, inspiration and learnings for the co-management stakeholders. Feedback loops in management integrate research and learning, without which, the initially high level of participation often fades. In joint discussions and dialogues, the co-management stakeholders give and receive knowledge in an integrated fashion, and ideas are linked with action. With participatory monitoring and evaluation, learnings and learning needs are collectively shared and analysed, creating common ideas, knowledge and experience.

12.7. Networking and Advocacy

The linkages developed by the community and its organizations to entities outside the community are crucial for the long-term success of co-management. Networking establishes linkages that are important so that community interests and concerns are taken into account by policy and legislative processes of government, to provide a source of technical assistance and knowledge, to achieve certain objectives and for sharing of experiences and strategies.

Many of these linkages were started early in the co-management programme as the community reached out to government and external agents. The linkages include those to:

- Other communities and projects with co-management;
- NGOs and business for technical assistance and information;
- Government agencies with which the community do not generally have contact;
- Sources of power and influence such as business leaders, politicians and law enforcement.

Advocacy is a mechanism through which organized groups and communities institutionalize their goals in policies and laws of other groups and government agencies. Advocacy is the act or process of supporting a cause or proposal. Advocacy can be undertaken at multiple levels including local, regional, national and international.

Advocacy requires organized action and it is always geared towards change, whether this is in terms of policies, perceptions or attitudes. Advocacy and networking sometimes need to go beyond merely sharing of information; in other cases these would involve making demands or claims on a particular issue. For example, a group of fishers may take a concerted action to stop the development of tourism in their area.

Generally, local advocacy work is simpler and requires fewer resources. Usually, the different stakeholders know each other and it is easier to immediately see the gains and setbacks experienced in the implementation of the advocacy plan. Advocacy revolves around concrete, identified issues that must be addressed. Priorities must be identified and focused on a few winnable issues at a time. Gaining some achievement in an issue builds the confidence of those involved and their morale.

National and international level advocacy work is much more complicated by virtue of its geographic and demographic scope. Advocacy at a national or international level revolves around legislative lobbying for reform.

A clear advocacy agenda is the starting point of the advocacy plan. The advocacy agenda defines the direction of what is wanted in policy changes. At the same time as the advocacy agenda is defined, the bottom-line negotiating positions should also be clarified. Once the advocacy agenda is defined, it has to be understood and explained to the constituency. This is where awareness-raising and social communication (see Chapter 8, Section 8.5) come in.

Advocacy is conducted internally to the community and through networking externally with other communities, community organizations, NGOs, local government, government agencies, academic and research institutions, donors, international agencies, projects and media. Advocacy is conducted by engaging government through its legislative, executive and judicial branches for legislative, policy and programme reform. The media, including the Internet, can be used to provide and disseminate information. Advocacy can also be undertaken through electoral engagement and metalegal forms such as pickets, rallies and demonstrations.

In both networking and advocacy, the strategy is to share information with other groups and communities and government so as to bring about greater understanding as well as social and policy changes. Maintaining dialogue and flexibility will facilitate better learning and sharing of information as the co-management programme progresses.

IV Post-implementation

13 'Turnover' or Post-implementation

Coastal Resources Institute, Prince of
Songkla University, Hat Yai.

The co-management programme is never completed. That is not to say that
plans and objectives are never achieved. Rather, as described earlier, co-
management is an iterative process of revisiting and revising plans and
activities, determining whether objectives have been achieved, adapting to
new conditions, and setting new objectives. It is not getting stuck in an endless
loop, but rather learning and adapting and moving forward. It is addressing
new changes and challenges.

Thus, as used here, 'turnover' or post-implementation does not mean
completing the co-management programme, but rather it means completing a

plan or activity. It means moving between a 'funded' programme and a self-sustaining community effort. It means evaluating the impacts of the plan or activity. It means making new plans. It means having the external agent and community organizer leave and turn over the co-management programme to the community. It means replication and extension of the co-management programme. It means 'scaling-up' the co-management programme to a larger geographic area or more issues. It means sustaining the programme. It means changing roles and responsibilities for the partners and stakeholders.

13.1. Turnover and Phase-out

External agents are often critical to facilitate the co-management programme. However, the external agent and the community should not view this as a permanent arrangement. Eventually, the co-management programme needs to be turned over to the community. The external agent should have a temporary relationship with the co-management programme, which may last several years, serving their particular function and then phasing out. This does not mean that the external agent should permanently leave the community. A bond has been established between the external agent and the community organizer. The community should be made aware that the external agent is there to provide assistance upon request. The external agent should have a phase-out plan that is well-understood by the community to eliminate any surprises and to minimize problems (Box 13.1). During this turnover and phase-out process, there will be changing roles and responsibilities of the stakeholders. Any agreements in effect should reflect these changes.

13.2. Post-evaluation

Every co-management programme should undertake a post-evaluation to determine how well it has achieved its goal and objectives. As mentioned in Chapter 10, Section 10.10, a summative or post-evaluation is undertaken after the programme's implementation where the focus is on a deeper analysis of results and outcomes and for determining the level of achievement of objectives and the impact of the programme. The results can be used for future planning and for drafting new or modifying agreements. This can be a link between the programme and the self-sustained community effort.

The post-evaluation effort should involve all stakeholders in meetings to discuss project results, hold general evaluation sessions, evaluate results against objectives, and plan for the next phase.

13.3. Scaling up

As previously discussed, the scale for co-management will vary a great deal depending on the community and the priority issues. In general, initially a

Box 13.1. Preparing for the Phase-out of a CBCRM Programme in Cabangan, Zambales, Philippines.

SIKAT is a Filipino NGO supporting CBCRM work in the province of Zambales in the Philippines. Specifically, they have been working closely with fisher organizations in implementing the *Fisheries Development and Management Programme* (FDMP). Over the past few years, it has assisted in the formation of municipality and provincial-wide federations of fisher organizations, implemented livelihood support activities, advocated changes in local and national policies, and provided overall support for capacity building for fishermen and women in Zambales.

Before SIKAT eventually phased out in the municipality of Cabangan in Zambales in 1999, it launched a mentoring project that included a series of capacity-building activities for selected community members. Called POGI in Filipino, meaning 'Pampamayanang Organisador para sa Gawaing Ikauunlad' (Local Community Organizers for Development Work), the mentoring project was premised on the belief that SIKAT and other external groups are merely facilitators of change, and that the communities should be able to act on their own in solving their own problems. SIKAT also believes that community organizations should be given opportunities to become independent of their partner NGO through a clear programme on mentoring and 'learning by doing' activities. In its phase-out, SIKAT did not sever its ties with community groups but re-invented its role as facilitators of change in the area.

In its initial phase-out stage, SIKAT facilitated the process of choosing local community organizers for this mentoring project. A total of 15 people, representing all community organizations, were chosen. The criteria used include the following:

- Active in organizational work;
- Respected and trusted by the organization;
- Open attitude in organizing and development work;
- Good health and physical condition;
- Committed to do the work related to the POGI project.

In its selection, the different community groups decided that there should be equal representation among women and men in the community.

The local community organizers were assigned to monitor and evaluate organizational activities, clarify processes and plan future activities. They also had successive trainings on:

- Training for trainers;
- Community organizing;
- Gender;
- Community microfinance;
- Community enterprise;
- Resource management;
- Issue advocacy;
- Participatory monitoring and evaluation;
- Financial management.

The trainings lasted about 15 months – the first 9 months were devoted to holding formal trainings and close mentoring and the last 6 months before phase-out was a 'trial' period for self-autonomy. During this time, SIKAT staff support were limited to monthly meetings with the community organizations.

During the initial months of the POGI programme, SIKAT staff were assigned as mentors of the local community organizers to closely sort out and apply learnings. Individually, the mentors monitored and assessed the activities of the local organizers. They got specific feedback on work assignments, assessed the difficult areas and got suggestions on how to overcome the difficulties. Monthly collective meetings were held where all the local organizers monitored activities, reflected on their experience and planned next steps.

POGI was implemented for 2 years. The first year involved capacity-building activities related to organizational strengthening. This included activities on project development and organizational systems, e.g. financial. The second year was devoted to launching more community-led activities. Only this year, the fisher organization did a research project on fisheries settlement problems, with minimal support from SIKAT.

Source: Balderrama and other SIKAT Mentors (1999).

limited scale (in terms of stakeholders, issues, jurisdiction) will support greater chances for success. Expansion of scale is often easier once initial activities succeed and are sustained. The issue of scaling up refers to the transferability of concepts, methods, and organizational and governance structures from one level to another in the dimensions of space and time. For example, scaling up from a single community to several communities to manage a bay-wide area through a co-management organization such as a bay-wide council which represents the different communities. Scaling up may also be undertaken to address issues of ecosystem-based management and broader social and economic links in the area. Questions related to scaling up include whether the same principles that guided the community-level co-management programme hold at the larger scale and what are the costs and benefits involved in scaling up and on community participation.

A number of factors constrain expansion including funding, resources, authority, management structure, voluntary basis of participation and others. Scaling up may be undertaken to include more stakeholder groups, manage a larger jurisdiction or management unit, and/or address new or a greater range of issues. In scaling up, a new plan and agreements will need to be developed. More information will need to collected and analysed. New groups may need to be organized. More funding will need to be sought. If the new scale involves multiple political jurisdictions, new legal support may be needed.

13.4. Replication and Extension

Every co-management programme has the potential to stimulate the start of co-management in other locations. Co-management in one community can contribute to the replication of the approach in other communities in a number of ways (Claridge, 1998):

- Providing an example which can be evaluated by members of other communities and government.
- Creating a group of community members who are experienced in the various aspects of process and methods and who can relate easily to members of other communities.
- Creating a precedent for involvement of government.
- Establishing new legislation and policies in support of co-management.
- Training people from other communities can create credibility for the original community and its efforts and ensure sustainability.
- Strengthen ties with local government and other concerned organizations, local, national and international.

Replication and extension does not necessarily have to wait until the post-implementation stage. As community members and project staff develop methodologies and solutions their confidence and capability improve. Once this stage is reached, these individuals can become advocates and trainers for the methodology or solution to other groups. Fisher-to-fisher training can be initiated. Members of other communities or government can be invited for a site visit to see what is being achieved. The external agent and its staff can initiate activities in other communities. Local government officials and staff can share their knowledge and experience with other local governments and with national government agencies.

It is clear that adopting replication and extension strategies can significantly multiply the benefits of an investment in co-management. Success often breeds success.

13.5. Sustainability

A number of factors have been identified as being critical for sustaining the co-management programme over time. These include, but are not limited to:

- Supportive policies and legislation are in place;
- Community organizations exist and are functioning;
- Community members have been empowered with skills and knowledge;
- Widespread public awareness and participation;
- There is widespread acceptance of the plan and agreements;
- Adequate financial resources and budget;
- Continual monitoring, evaluation and adaptation of the plan;
- Supportive local government;
- High level of compliance with management measures;
- Economic benefits are being received by the community.

V Conclusion

14 Making It Happen!

In this handbook, we have presented you with a very comprehensive and detailed reference on 'a' process for community-based co-management. It is a 'generic' process not 'the' process. As is hopefully clear at this point, there is no 'the' process since each situation where co-management may be used will be different. How co-management is undertaken in South Africa is different than in Belize and in the Philippines. There is a range of activities and methods that can be used in planning and implementing a co-management programme. This handbook is meant to stimulate thought and action. The handbook provides ideas, methods, techniques, activities, checklists, examples, questions and indicators for planning and implementation. It is meant to serve as a source of information for the practical planning and implementation of a co-management programme.

The user should become familiar with the complete process as presented in this handbook, think about the situation where co-management will be considered for use, and adapt the co-management process to the situation or community.

As presented in this handbook, there are a great many activities involved in undertaking a co-management programme. Some or all of the activities and methods may be relevant. It can be intimidating and overwhelming to think about what to do and where to start. After reviewing this handbook, you might say to yourself that you or your organization cannot possibly do all this. It would take too much money and time and effort. You don't have to do it all or do it all at once. If you do, you might possibly fail. It is best to start simple and keep it simple. Assess where the community is in the process and what needs to be done first – identify the problem and issues? Organize the fishing community? Empower the fishers with environmental education? Set priorities and move forward. Use the resources which are available to you.

The concept of community-based co-management has gained acceptance among governments, development agencies and development practitioners as an alternative fisheries management strategy to the top-down, centralized government approach to fisheries management. However, co-management may not be an appropriate alternative management strategy for every community or area. Centralized management may be more appropriate to the community or area. The development of co-management is neither automatic nor simple, nor is its sustainability guaranteed. Co-management does show promise for addressing many of the requirements for sustainability, equity and efficiency in fisheries and coastal resources management.

As presented in the final paragraphs of the book *Managing Small-scale Fisheries: Alternative Directions and Methods* (Berkes *et al.*, 2001):

> The reality is that we have to do something. Most small-scale fisheries and fishers around the world are in crisis. Since the current management approaches are not effective, trying something new may be better than maintaining the status quo.

What do you have to do? You will have to learn and think about these new concepts and techniques. Yes, it will take some work. It will involve study and discussion with others. As difficult as it may be, you will have to put aside your biases, whatever they may be about: the behaviour of fishers, the behaviour of managers, scientific superiority, the corruption of government. If we are to succeed, we must open our minds and refresh our thinking. This sounds transcendental; new directions often are. But what choice do we have? The future of our marine and coastal resources is at stake. People's lives and futures are at stake. **You can make a difference!**

References

Abad, G.S. 1997. Community organizing in the fisheries sector program: lessons learned. *Tambuli* 2, 7–10.

Aleroza, R., O. Fraxedez, A. Bautista and R. Barangas. 2003. *Session, Siesta at Socials: Paghahalaw ng Karanasan, Pananaw at mga Aral sa Pag-oorganisa sa CBCRM.* A.M.M. Dasig (ed.) Quezon City: CBCRM Resource Center, UP Social Action and Research Development Foundation, Inc. and UP College of Social Work and Community Development, pp. 19–42.

Almerigi, S. 2000. *Leadership for Fisherfolk.* CARICOM Fishery Research Document No. 24. CARICOM Fisheries Unit, Belize City, Belize.

Arciaga, O., F. Gervacio, R.C. Capistrano and C. Demesa. 2002. *Envisioning Life: Community-Created Sustainable Livelihood Analysis and Development.* Graham, J., C.M. Nozawa and A. Plantilla (eds) Philippines: Haribon Foundation with support from the International Development Research Centre, 62 pp.

Balderrama, D. and other SIKAT Mentors. 1999. *POGI: Pampamayanang Organisador para sa Gawaing Ikauunlad: A Concept Paper on Local Community Organizers/ Development Workers.* Internal programme document.

Beddington, J.R. and R.B. Rettig. 1983. *Approaches to the Regulation of Fishing Effort.* Fisheries Technical Report No. 243. Food and Agriculture Organization of the United Nations, Rome.

Berkes, F., R. Mahon, P. McKinney, R. Pollnac and R. Pomeroy. 2001. *Managing Small-scale Fisheries: Alternative Directions and Methods.* International Development Research Centre, Ottawa, Canada.

Blaxter, J.H.S. 2000. The enhancement of marine fish stocks. *Advances in Marine Biology* 38, 1–58.

Borrini-Feyerabend, G. 1996. *Collaborative Management of Protected Areas: Tailoring the Approach to the Context.* IUCN, Gland, Switzerland.

Borrini-Feyerabend, G. (ed.) 1997. *Beyond Fences: Seeking Social Sustainability in Conservation*, 2 Volumes. IUCN, Gland, Switzerland.

Borrini-Feyerabend, G., M. Taghi Farvar, J.C. Nguinguiri and V. Ndangang. 2000. *Co-management of Natural Resources: Organizing, Negotiating and Learning-by-doing.* GTZ and IUCN, Kasparek Verlag, Heidelberg, Germany.

Borrini-Feyerabend, G., M. Pimbert, M.T. Farvar, A. Kothari and Y. Renard. 2004. *Sharing Power: Learning by Doing in Co-management of Natural Resources Throughout the World.* International Institute for Environment and Development, World Conservation Bookstore, IUCN-

Collaborative Management Working Group, Cenesta, Tehran, Iran.

Bromley, D. 1991. *Environment and Economy: Property Rights and Public Policy*. Blackwell, Cambridge, Massachusetts.

Brown, N. 1997. *Devolution of Authority Over the Management of Natural Resources: The Soufriere Marine Management Area, St. Lucia*. Caribbean Centre for Development Administration and Caribbean Natural Resources Institute, Trinidad.

Brzeski, V., J. Graham and G. Newkirk. 2001. *Participatory Research and CBCRM: In Context*. Coastal Resources Research Network, Dalhousie University, Halifax, Nova Scotia, Canada and International Development Research Centre, Ottawa, Canada.

Buckles, D. and G. Rusnak. 1999. Introduction: conflict and collaboration in natural resource management. In: Buckles, D. (ed.) *Cultivating Peace: Conflict and Collaboration in Natural Resource Management*. International Development Research Centre, Ottawa, Canada.

Buhat, D.Y. 1994. Community-based coral reef and fisheries management, San Salvador Island, Philippines. In: White, A., L.Z. Hale, Y. Renard and L. Cortesi (eds) *Collaborative and Community-based Management of Coral Reefs*. Kumarian Press, West Hartford, Connecticut.

Bunce, L., P. Townsley, R. Pomeroy and R. Pollnac. 2000. *Socioeconomic Manual for Coral Reef Management*. Australian Institute of Marine Sciences, Townsville, Australia.

Caddy, J. 1999. Fisheries management in the twenty-first century: will new paradigms apply? *Reviews in Fish Biology* 9, 1–43.

Cambodia Family Development Services, unpublished programme documents. Rivera-Guieb, R. 2004. A Report on the Training Needs Analysis on Community Fisheries for the Cambodia Family Development Services Staff Working in Pursat Province. A report for CFDS with support from Oxfam-America. Phnom Penh, Cambodia.

Casia, M. 2000. *Introduction to the Establishment of a Community-based Marine Sanctuary*. CRM Document No. 24–CRM/2000.

Coastal Resources Management Project, Cebu City, Philippines.

Center for Rural Progress and the International Marinelife Alliance-Vietnam. 2003. *Community-based Coastal Resource Management: Capacity in the Vietnam Context*. Hanoi.

Chambers, R. 1994. Participatory rural appraisal (PRA): analysis of experience. *World Development* 22, 1253–1268.

Chapman, M.D. 1987. Traditional political structure and conservation in Oceania. *Ambio* 16, 201–205.

Chevalier, J.M. and D. Buckles. 1999. Conflict management: a heterocultural perspective. In: Buckles, D. (ed.) *Cultivating Peace: Conflict and Collaboration in Natural Resource Management*. International Development Research Centre, Ottawa, Canada.

Christie, P. and A. White. 1997. Trends in development of coastal area management in tropical countries: from central to community orientation. *Coastal Management* 25, 155–181.

Claridge, G. 1998. Designing and implementing a co-management approach. In: Claridge, G. and B. O'Callaghan (eds) *Community Participation in Wetlands Management: Lessons From the Field*. Wetlands International, Wageningen, Netherlands.

Cleofe, M.J.T. 1999. Mainstreaming gender. In: Oxfam-Great Britain and Partners (eds) *Gleanings: Lessons in Community-based Coastal Resources Management*. Oxfam-Great Britain: Philippines Office, Quezon City, Philippines.

Commission of the European Communities. 1993. *Project Cycle Management, Integrated Approach and Logical Framework*. Commission of the European Communities, Brussels, Belgium.

Community-based Forest Resource Conflict Management: A Training Package. 2002. Food and Agriculture Organization of the United Nations, Rome and Regional Community Forestry Centre, Bangkok, Thailand.

Copes, P. 1986. A critical review of the individual quota as a device in fisheries management. *Land Economics* 62, 278–291.

Dayton, P.K., S.F. Thrush, M.T. Agardy and R.J.

Hofman. 1995. Environmental effects of marine fishing. *Aquatic Conservation: Marine and Freshwater Ecosystems* 5, 205–232.

Department of Environment and Natural Resources, Bureau of Fisheries and Aquatic Resources of the Department of Agriculture, and Department of the Interior and Local Government. 2001a. *Philippine Coastal Management Guidebook No. 2: Legal and Jurisdictional Framework for Coastal Management*. Coastal Resource Management Project of the Department of Environment and Natural Resources. Cebu City, Philippines.

Department of Environment and Natural Resources, Bureau of Fisheries and Aquatic Resources of the Department of Agriculture, and Department of the Interior and Local Government. 2001b. *Philippine Coastal Management Guidebook No. 3: Coastal Resource Management Planning*. Coastal Resource Management Project of the Department of Environment and Natural Resources, Cebu City, Philippines.

Department of Environment and Natural Resources, Bureau of Fisheries and Aquatic Resources of the Department of Agriculture, and Department of the Interior and Local Government. 2001c. *Philippine Coastal Management Guidebook No. 4: Involving Communities in Coastal Management*. Coastal Resource Management Project of the Department of Environment and Natural Resources, Cebu City, Philippines.

Department of Environment and Natural Resources, Bureau of Fisheries and Aquatic Resources of the Department of Agriculture, and Department of the Interior and Local Government. 2001d. *Philippine Coastal Management Guidebook No. 8: Coastal Law Enforcement*. Coastal Resource Management Project of the Department of Environment and Natural Resources, Cebu City, Philippines.

Edmunds, D. and E. Wollenberg. 2002. *Disadvantaged Groups in Multistakeholder Negotiations*. CIFOR Programme Report. Center for International Forestry Research, Bogor, Indonesia.

English, S., C. Wilkinson and V. Barber. 1994. *Survey Manual for Tropical Marine Resources*, 2nd Edition. Australian Institute of Marine Sciences, Townsville, Australia.

Espeut, P. nd. *A Process Towards Co-management*. Caribbean Coastal Area Management Foundation, Clarendon, Jamaica.

Fellizar, F. 1994. Achieving sustainable development through community-based management. In: Siason, I. and R. Subalde (eds) *Community-based Management of Coastal Resources*. University of the Philippines in the Visayas, Iloilo City, Philippines.

Ferrer, E. and C. Nozawa. 1997. *Community-based Coastal Resource Management in the Philippines: Key Concepts, Methods and Lessons Learned*. A paper presented at the International Development Research Centre planning workshop on community-based natural resource management, 12–16 May, Hue, Vietnam.

Fogarty, M. 1999. Essential habitat, marine reserves and fishery management. *Trends in Ecology and Evolution* 14, 133–134.

Food and Agriculture Organization. 1996. *Integration of Fisheries into Coastal Area Management*. Fisheries Department, FAO, Rome.

Freeman, M.M.R., Y. Matsuda and K. Ruddle (eds) 1991. Adaptive marine resource management systems in the Pacific. *Resource Management and Optimization* 8, 127–245.

Galit, J. 2001. Catching power: a story of Honda Bay CBCRM. In: Ferrer, E.M., L.P. de la Cruz and G.F. Newkirk (eds) *Hope Takes Root: Community-based Coastal Resources Management Stories in Southeast Asia*. CBCRM Learning Center, Quezon City, Philippines and Coastal Resources Research Network, Dalhousie University, Halifax, Nova Scotia, Canada.

Gibbs, C. and D. Bromley. 1987. *Institutional Development for Local Management of Rural Resources*. East-West Environment and Policy Institute, Workshop Report No. 2, Honolulu, Hawaii.

Green, S. 1997. *Public Awareness and Education Campaigns for MPAs*. Coastal Resources Management Project, Cebu City, Philippines.

Grimble, R. and M.-K. Chan. 1995. Stakeholder analysis for natural resource management in developing countries. *Natural Resources Forum* 19, 113–124.

Hale, L.Z. and M. Lemay. 1994. Coral reef protection in Phuket, Thailand: a step toward integrated coastal management. In: White. A., L.Z. Hale, Y. Renard and L. Cortesi (eds) *Collaborative and Community-based Management of Coral Reefs.* Kumarian Press, West Hartford, Connecticut.

Hara, M., S. Donda and F.J. Njaya. 2002. Lessons from Malawi's experience with fisheries co-management initiatives. In: Geheb, K. and M.-T. Sarch (eds) *Africa's Inland Fisheries: The Management Challenge.* Fountain Publishers, Kampala, Uganda.

Harris, J., G. Branch, C. Sibiya and C. Bill. 2003. The Sokhulu subsistence mussel-harvesting project: co-management in action. In: Hauck, M. and M. Sowman (eds) *Waves of Change: Coastal and Fisheries Co-management in Southern Africa,* Chapter 4. University of Cape Town Press, Landsdowne, South Africa.

Hauck, M. and R. Hector. 2003. Towards abalone and rock lobster co-management in the Hangklip-Kleinmond area. In: Hauck, M. and M. Sowman (eds) *Waves of Change: Coastal and Fisheries Co-management in Southern Africa,* Chapter 11. University of Cape Town Press, Landsdowne, South Africa.

Heinen, A. 2003. *Rehabilitating Nearshore Fisheries: Theory and Practice on Community-based Coastal Resources Management from Danao Bay, Philippines.* Community Based Coastal Resources Management Center, UP Social Action Research and Development Foundation, UP College of Social Work and Community Development, Oxfam Great Britain, SNV-Philippines, and Pipil Foundation, Quezon City, Philippines.

Hilborn, R. and C. Walters. 1992. *Quantitative Fisheries Stock Assessment.* Chapman and Hall, London, UK.

Horton, D. 2002. *Planning, Implementing, and Evaluating Capacity Development.* Briefing Paper No. 50. International Service for National Agricultural Research, The Hague, Netherlands.

Hviding, E. and G. Baines. 1996. Custom and complexity: marine tenure, fisheries management and conservation in Marovo Lagoon, Solomon Islands. In: Howitt, R. *et al.* (eds) *Resources, Nations and Indigenous Peoples: Case Studies from Australasia, Melanesia and Southeast Asia.* Oxford University Press, Australia.

Inter-American Development Bank. 1997. Evaluation – a management tool for improving project performance. *Logical Framework Analysis.* IADB, Washington, DC.

International Institute of Rural Reconstruction. 1998. *Participatory Methods in Community-based Coastal Resource Management.* 3 volumes. Silang, Cavite, Philippines.

International Marinelife Alliance Vietnam. nd. *Final Report on Trao Reef Locally Managed Marine Reserve.* Hanoi, Vietnam.

Jackson, B. and A. Ingles. 1995. *Participatory Techniques for Community Forestry: A Field Manual.* Nepal-Australia Community Forestry Project. Technical Note 5/95. ANUTECH Pty Ltd, Canberra.

Johannes, R. 1978. Traditional marine conservation methods in Oceania and their demise. *Annual Reviews of Ecology and Systematics* 9, 349–364.

Johannes, R. 1998. Government-supported, village-based management of marine resources in Vanuatu. *Ocean and Coastal Management* 40, 243–246.

Johnson, M. 1992. *Lore: Capturing Traditional Environmental Knowledge.* International Development Research Centre, Ottawa, Canada.

Juinio-Menez, M.A., S. Salmo, E. Tamayo, N. Estepa, H. Bangi and P. Alino. 2000. *'Bugsay': Community Environmental Education: Experiences From Bolinao, Northern Philippines.* Community-based coastal resources management program, Marine Science Institute, University of the Philippines, Dilliman, Quezon City, Philippines.

Kapetsky, J.M. 1985. *Some Considerations for the Management of Coastal Lagoon and Estuarine Fisheries.* Fisheries Technical Paper No. 218. Food and Agriculture Organization of the United Nations, Rome.

Katon, B., R. Pomeroy, L. Graces and A. Salamanca. 1999. Fisheries Management of San Salvador Island, Philippines: A Shared Responsibility. *Society and Natural Resources* 12, 777–795.

Khai Syrado and Thai Kimseng. 2002. *Community Participation in Fisheries Management*. Department of Fisheries, Phnom Penh, Cambodia.

Kolter, P. *et al*. 2002. *Social Marketing: Improving the Quality of Life*, 2nd Edition. Sage Publications, Thousand Oaks, California.

Korten, D. (ed.) 1987. *Community Management: Asian Experiences and Perspectives*. Kumarian Press, West Hartford, Connecticut.

Krishnarayan, V., T. Geoghegan and Y. Renard. 2002. *Assessing Capacity for Participatory Natural Resources Management*. Caribbean Natural Resources Institute, Trinidad.

Landon, S. and S. Langill. 1998. *Participatory Research: Readings and Resources for Community-based Natural Resource Management Researchers*, Volume 3. International Development Research Centre, Ottawa, Canada.

Langill, S. and S. Landon. 1998. *Indigenous Knowledge* – Volume 4. Community-based natural resource management program initiative, International Development Research Centre, Ottawa, Canada.

Lemay, M. and L.Z. Hale. 1989. *Coastal Resources Management: A Guide to Public Education Programs and Materials*. Kumarian Press, West Hartford, Connecticut.

Maine, R.A., B. Cam and D. Davis-Case. 1996. *Participatory Analysis, Monitoring and Evaluation for Fishing Communities: A Manual*. Food and Agriculture Organization of the United Nations, Rome.

Mangahas, M.F. 1993. Mataw-amung nu rayon, Anito. Man, the Fish of Summer, and the Spirits: An Ethnography of Mataw Fishing in Batanes. Unpublished masteral thesis, University of the Philippines, Diliman, Quezon City, Philippines.

Margoluis, R. and N. Salafsky. 1998. *Measures of Success: Designing, Managing and Monitoring Conservation and Development Projects*. Island Press, Washington, DC.

McCay, B. and S. Jentoft. 1996. From the bottom up: participatory issues in fisheries management. *Society and Natural Resources* 9, 237–250.

McConney, P., R. Mahon and H. Oxenford. 2003a. *Barbados Case Study: The Fisheries Advisory Committee*. Caribbean Conservation Association and Centre for Resource Management and Environmental Studies–University of the West Indies, Barbados.

McConney, P., R. Pomeroy and R. Mahon. 2003b. *Guidelines for Coastal Resource Co-management in the Caribbean: Communicating the Concepts and Conditions That Favour Success*. Caribbean Conservation Association and Centre for Resource Management and Environmental Studies–University of the West Indies, Barbados.

Moore, C. 1986. *The Mediation Process: Practical Strategies for Managing Conflict*, 1st edn. Jossey-Bass, San Francisco, California.

Moore, C. 1996. *The Mediation Process: Practical Strategies for Managing Conflict*, 2nd edn. Jossey-Boss, San Francisco, California.

Narayan, D. 1996. What is participatory research? In: *Toward Participatory Research*. World Bank, Washington, DC, pp. 17–30.

National Roundtable on the Environment and the Economy. 1998. *Sustainable Strategies for Oceans: A Co-management Guide*. Ottawa, Canada.

Neis, B., D. Schneider, L. Felt, R. Haedrich, J. Fischer and J. Hutchings. 1999. Fisheries assessments: what can be learned from interviewing resource users. *Canadian Journal of Fisheries and Aquatic Sciences* 56, 1949–1963.

Nielsen, J. and T. Vedsmand. 1995. User participation and institutional change in fisheries management: a viable alternative to the failures of 'top-down' driven control? *Ocean and Coastal Management* 42, 19–47.

Njaya, F.J. 2002. Fisheries co-management in Malawi: implementation arrangements for Lake Malombe, Chilwa and Chuita. In: Geheb, K. and M.-T. Sarch (eds) *Africa's Inland Fisheries: The Management Challenge*. Fountain Publishers, Kampala, Uganda.

Nong, Kim, Ouk Ly Khim and Khy An. 2004. *Community Organizing: Working to Create and Support Village Management Committees.* A paper presented at the International Association of Common Property Tenth Biennial Conference in Oaxaca, Mexico, 9–13 August 2004.

North, D.C. 1990. *Institutions, Institutional Change and Economic Performance.* Cambridge Series of Political Economy of Institutions and Decisions. Cambridge University Press, Cambridge, UK.

Oposa, A.A. 1996. Legal marketing of environmental law. *Duke Journal of Comparative and International Law* 6, 273–291.

Ostrom, E. 1990. *Governing the Commons: The Evolution of Institutions for Collective Action.* Cambridge University Press, Cambridge, UK.

Ostrom, E. 1991. *A Framework for Institutional Analysis.* Working paper. Indiana University, Workshop in Political Theory and Policy Analysis.

Ostrom, E. 1992. *Crafting Institutions for Self-governing Irrigation Systems.* Institute for Contemporary Studies, San Francisco, California.

Pendzich, C., G. Thomas and T. Wohlgenant. 1994. *The Role of Alternative Conflict Management in Community Forestry.* Food and Agriculture Organization of the United Nations, Rome.

Phap, Tôn Thât. 2002. Co-management in the planning of a waterway system for aquaculture. In: Brezski, V. and G.F. Newkirk (eds) *Lessons in Resource Management from the Tam Giang Lagoon.* CoRR, CIDA and IDRC, Halifax, Nova Scotia, Canada, pp. 39–52.

Piquero, D. 2004. *Creation of Bohol Coastal Law Enforcement Councils.* A paper presented at the Regional Workshop on Governance in Community-Based Coastal Resource Management: Experiences and Lessons in Participation. Organized by the CBCRM Learn Network and held in Binangonan, Rizal, Philippines on 1–4 March 2004.

Pollnac, R. and J.J. Poggie. 1997. *Fish Aggregating Devices in Developing Countries: Problems and Perspectives.* International Center for Marine Resources Development and Anthropology Program, University of Rhode Island, Kingston, Rhode Island.

Pollnac, R.B., R.S. Pomeroy and L. Bunce. 2003. Factors Influencing the Sustainability of Integrated Coastal Management Projects in Central Java and North Sulawesi, Indonesia. *Indonesian Journal of Coastal and Marine Resources* Special Edition 1, 24–33.

Pomeroy, R. and T. Goetze. 2003. *Belize Case Study: Marine Protected Areas Co-managed by Friends of Nature.* Caribbean Conservation Association and Centre for Resource Management and Environmental Studies-University of the West Indies, Barbados.

Pomeroy, R., B. Katon and I. Harkes. 2001. Conditions affecting the success of fisheries co-management: lessons from Asia. *Marine Policy* 25, 197–208.

Rijsberman, F. (ed.) 1999. *Conflict Management and Consensus Building for Integrated Coastal Management in Latin America and the Caribbean.* ENV-132. Inter-American Development Bank, Washington, DC.

Rivera, R. 1997. *Re-inventing Power and Politics in Coastal Communities: Community-based and Coastal Resource Management in the Philippines.* Marine Affairs Program, Dalhousie University, Nova Scotia, Canada.

Rivera-Guieb, R. and M. Marschke. 2002. *A Field Manual for Facilitators on the Basic Concept and Process of Community Fisheries Management.* Department of Fisheries, Phnom Penh, Cambodia.

Ruddle, K. 1994. Changing the focus of coastal fisheries management. In: R.S. Pomeroy (ed.) *Community Management and Common Property of Coastal Fisheries in Asia and the Pacific: Concepts, Methods and Experiences.* ICLARM Proceedings 45. International Center for Living Aquatic Resources Management, Manila, Philippines.

Ruddle, K. and T. Akimichi. 1984. *Maritime Institutions in the Western Pacific.* National Museum of Ethnology, Senri Ethnological Studies 17, Osaka, Japan.

Runge, C.F. 1992. Common Property and Collective Action in Economic Development. In: D. Bromley *et al.* (eds) *Making the Commons Work: Theory, Practice, and Policy.* ICS Press, San Francisco, California.

Sajise, P. 1995. *Community-based Resource Management in the Philippines: Perspectives and Experiences.* A paper presented at the Fisheries Co-management workshop, North Sea Center, 29–31 May, Hirtshals, Denmark.

Sen, S. and J.R. Nielsen. 1996. Fisheries co-management: a comparative analysis. *Marine Policy* 20, 405–418.

Sowman, M. 2003. Co-management of the Olifants River harder fishery. In: Hauck, M. and M. Sowman (eds) *Waves of Change: Coastal and Fisheries Co-management in Southern Africa,* Chapter 12. University of Cape Town Press, Landsdowne, South Africa.

Sowman, M., M. Hauck and G. Branch. 2003. Lessons learned from nine coastal and fisheries co-management case studies. In: Hauck, M. and M. Sowman (eds) *Waves of Change: Coastal and Fisheries Co-management in Southern Africa,* Chapter 13. University of Cape Town Press, Landsdowne, South Africa.

Spergel, B. and M. Moye. 2004. *Financing Marine Conservation: A Menu of Options.* World Wildlife Fund Center for Conservation Finance, Washington, DC.

Sverdrup-Jensen, S. and J.R. Nielsen. 1998. Co-management in small-scale fisheries: a synthesis of southern and west Africa experiences. In: Normann, A., J.R. Nielsen and S. Sverdrup-Jensen (eds) *Fisheries Co-management in Africa: Proceedings from a Regional Workshop on Fisheries Co-management Research.* Institute for Fisheries Management and Coastal Community Development, Hirtshals, Denmark.

Tambuyog Development Center and Center for Empowerment and Resource Development. 1998. *CBCRM School: Training modules.* Quezon City, Philippines.

Tang, S.Y. 1992. *Institutions and Collective Action: Self-governance in Irrigation.* ICS Press, San Francisco, California.

Tanyang, G. 2001. *Women in Fisheries: A Review of Limited Literature.* A paper commissioned by the NGOs for Fisheries Reform for the Workshop on Fisheries and Gender held on 25–26 September 2001.

Townsend, R. and J. Wilson. 1987. An Economic View of the Tragedy of the Commons; From Privatization to Switching. In: Acheson, J. and B. McCay (eds) *The Question of the Commons.* University of Arizona Press, Tucson, Arizona.

Townsley, P. 1996. *Rapid Rural Appraisal, Participatory Rural Appraisal and Aquaculture.* FAO Fisheries Technical Paper no. 358. Food and Agriculture Organization of the United Nations, Rome.

UNDP. 1998. *Capacity Assessment and Development in a Systems and Strategic Management Context.* Management Development and Governance Division, Technical Advisory Paper No. 3. United Nations Development Program, New York.

United States Agency for International Development. 1994. *The Logical Framework: A Project Level Design Tool.* Document from a videotape presentation produced by the Professional Studies and Career Development Division, USAID and PASITAM of the Midwest Universities Consortium for International Activities, Washington, DC.

Vernooy, R. 1999. *Participatory Monitoring and Evaluation: Readings and Resources for Community-based Natural Resource Management Researchers.* Volume 8. International Development Research Centre, Ottawa, Canada.

Volunteer Service Overseas–Philippines. nd. *Managing Tensions and Conflicts Over Natural Resources.* Quezon City, Philippines.

Wallerstein, N. 1992. Powerlessness, empowerment, and health: implications for health promotion programs. *American Journal of Health Promotion* 6, 197–205.

Walters, J.S., J. Maragos, S. Siar and A. White. 1998. *Participatory Coastal Resource Assessment: A Handbook for Community Workers and Coastal Resource Managers.* Coastal Resource Management Project and Silliman University, Cebu City, Philippines.

Warner, M. 2001. *Complex Problems, Negotiated Solution: Tools to Reduce Conflict in Community Development.* ITDG Publishing, UK.

Welcomme, R.L. 1998. Evaluation of stocking and introductions as management tools. In: Cowx, I.G. (ed.) *Stocking and Introduction of Fish.* Fishing News Books, London, pp. 397–413.

White, A. 1988. *Marine Parks and Reserves: Management for Coastal Environments in Southeast Asia.* ICLARM Education Series 2. International Center for Living Aquatic Resources Management. Manila, Philippines.

White, A., L.Z. Hale, Y. Renard and L. Cortesi. 1994. *Collaborative and Community-based Management of Coral Reefs.* Kumarian Press, West Hartford, Connecticut.

Wilson, J.A., J.M. Acheson, M. Metcalfe and P. Kleban. 1994. Chaos, complexity and community management of fisheries. *Marine Policy* 184, 291–305.

Further Reading

Andrews, E. 1998. *Supporting Community Based Environmental Education.* Discussion Paper Summary. US Environmental Protection Agency, Region 10.

Bay of Bengal Programme. 1993. *A Manual on Rapid Appraisal Methods for Coastal Communities.* Bay of Bengal Programme, Madras, India.

Brown, J., A. Kothari and M. Menon. 2002. Local communities and protected areas. *PARKS* Vol. 12, No. 2. IUCN, Gland, Switzerland.

Co, E. 1999. *Beating the Drums: Advocacy for Policy Reform in the Philippines.* Oxfam Great Britain–Philippines, Quezon City.

Community Tool Box. http://ctb.ku. edu/tools/en/ (Building leadership, organizational structure, choosing a group)

Community Organizing. http://www.marin institute.org/action_packs/community_org. htm

Community Organizing Toolbox. http:// www.nfg.org/cotb/

Fisheries Management in Community-based Coastal Resource Management, Volume 1. 2003. Oxfam-Great Britain and the Community based Coastal Resource Management Resource Center, Quezon City, Philippines.

Hess, D. nd. *Community Organizing, Building and Developing: Their Relationship to Comprehensive Community Initiatives.* http://comm-org.utoledo.edu/ papers99/hess3.htm

International Institute for Environment and Development. 1997. *Participation and Fishing Communities.* PLA Notes 30. London.

Kaner, S. with L. Lind, C. Toldi, S. Fisk and D. Berger. 1996. *Facilitators Guide to Participatory Decision-making.* New Society Publishers, Gabriola Island, BC, Canada.

Mahon, R., S. Almerigi, P. McConney, C. Parker and L. Brewster. 2003. Participatory methodology used for sea urchin co-management in Barbados. *Ocean and Coastal Management* 46, 1–25.

Milne, N., P. Christie, R. Oram, R.-L. Eisma and A. White. 2003. *Integrated Coastal Management Process Sustainability Reference Book.* University of Washington School of Marine Affairs, Silliman University, the Coastal Resource Management Project, Cebu City, Philippines.

Neiland, A.E. and C. Bene (eds) 2004. *Poverty and Small-scale Fisheries in West Africa.* Kluwer Academic, Dordrecht, Netherlands.

Participatory Monitoring and Evaluation in Community-based Coastal Resource Management, Volume 3. 2003. Oxfam Great Britain and the Community based Coastal Resource Management Resource Center, Quezon City, Philippines.

Pido, M., R. Pomeroy, M. Carlos and L. Garces. 1996. *A Handbook for Rapid Appraisal of Fisheries Management Systems.* International Center for Living Aquatic Resources Management, Manila, Philippines.

Pollnac, R.B. 1989. *Monitoring and Evaluating the Impacts of Small-scale Fishery Projects.* International Center for Marine Resource

Development, University of Rhode Island, Kingston, Rhode Island.

Sustainable Livelihoods in Community-based Coastal Resource Management, Volume 2. 2003. Oxfam Great Britain and the Community based Coastal Resource Management Resource Center, Quezon City, Philippines.

Viswanathan, K., J.R. Nielsen, P. Degnbol, M. Ahmed, M. Hara and N. Mustapha. 2003. *Fisheries Co-management Policy Brief: Findings from a Worldwide Study*. World Fish Center, Penang, Malaysia.

Wiber, M., F. Berkes, A. Charles and J. Kearney. 2004. Participatory research supporting community-based fishery management. *Marine Policy* 28, 459–468.

Index

Note: page numbers in *italics* refer to tables and boxes.

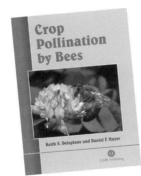
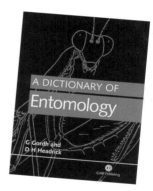
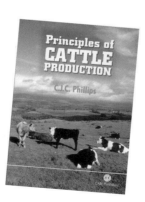